Ecosystems and Human Well-being

Synthesis

A Report of the Millennium Ecosystem Assessment

Core Writing Team

Walter V. Reid, Harold A. Mooney, Angela Cropper, Doris Capistrano, Stephen R. Carpenter, Kanchan Chopra, Partha Dasgupta, Thomas Dietz, Anantha Kumar Duraiappah, Rashid Hassan, Roger Kasperson, Rik Leemans, Robert M. May, Tony (A.J.) McMichael, Prabhu Pingali, Cristián Samper, Robert Scholes, Robert T. Watson, A.H. Zakri, Zhao Shidong, Neville J. Ash, Elena Bennett, Pushpam Kumar, Marcus J. Lee, Ciara Raudsepp-Hearne, Henk Simons, Jillian Thonell, and Monika B. Zurek

Extended Writing Team

MA Coordinating Lead Authors, Lead Authors, Contributing Authors, and Sub-global Assessment Coordinators

Review Editors

José Sarukhán and Anne Whyte (co-chairs) and MA Board of Review Editors

Suggested citation:

Millennium Ecosystem Assessment, 2005. *Ecosystems and Human Well-being: Synthesis*.
Island Press, Washington, DC.

ISLAND PRESS is a trademark of The Center for Resource Economics.

Library of Congress Cataloging-in-Publication data.

Ecosystems and human well-being : synthesis / Millennium Ecosystem Assessment.
p. cm. – (The Millennium Ecosystem Assessment series)
ISBN 1-59726-040-1 (pbk. : alk. paper)
1. Human ecology. 2. Ecosystem management. I. Millennium Ecosystem Assessment (Program) II. Series.
GF50.E26 2005
304.2–dc22
2005010265

British Cataloguing-in-Publication data available.

Printed on recycled, acid-free paper

Book design by Dever Designs

Manufactured in the United States of America

CONTENTS

Foreword   ii

Preface   v

Reader's Guide   x

Summary for Decision-makers   1

    Finding 1: Ecosystem Change in Last 50 Years   2

    Finding 2: Gains and Losses from Ecosystem Change   5

    Finding 3: Ecosystem Prospects for Next 50 Years   14

    Finding 4: Reversing Ecosystem Degradation   18

Key Questions in the Millennium Ecosystem Assessment   25

    1. How have ecosystems changed?   26

    2. How have ecosystem services and their uses changed?   39

    3. How have ecosystem changes affected human well-being and poverty alleviation?   49

    4. What are the most critical factors causing ecosystem changes?   64

    5. How might ecosystems and their services change in the future under various plausible scenarios?   71

    6. What can be learned about the consequences of ecosystem change for human well-being at sub-global scales?   84

    7. What is known about time scales, inertia, and the risk of nonlinear changes in ecosystems?   88

    8. What options exist to manage ecosystems sustainably?   92

    9. What are the most important uncertainties hindering decision-making concerning ecosystems?   101

Appendix A. Ecosystem Service Reports   103

Appendix B. Effectiveness of Assessed Responses   123

Appendix C. Authors, Coordinators, and Review Editors   132

Appendix D. Abbreviations, Acronyms, and Figure Sources   136

Appendix E. Assessment Report Tables of Contents   137

FOREWORD

The Millennium Ecosystem Assessment was called for by United Nations Secretary-General Kofi Annan in 2000 in his report to the UN General Assembly, *We the Peoples: The Role of the United Nations in the 21st Century.* Governments subsequently supported the establishment of the assessment through decisions taken by three international conventions, and the MA was initiated in 2001. The MA was conducted under the auspices of the United Nations, with the secretariat coordinated by the United Nations Environment Programme, and it was governed by a multistake-holder board that included representatives of international institutions, governments, business, NGOs, and indigenous peoples. The objective of the MA was to assess the consequences of ecosystem change for human well-being and to establish the scientific basis for actions needed to enhance the conservation and sustainable use of ecosystems and their contributions to human well-being.

This report presents a synthesis and integration of the findings of the four MA Working Groups (Condition and Trends, Scenarios, Responses, and Sub-global Assessments). It does not, however, provide a comprehensive summary of each Working Group report, and readers are encouraged to also review the findings of these separately. This synthesis is organized around the core questions originally posed to the assessment: How have ecosystems and their services changed? What has caused these changes? How have these changes affected human well-being? How might ecosystems change in the future and what are the implications for human well-being? And what options exist to enhance the con-servation of ecosystems and their contribution to human well-being?

This assessment would not have been possible without the extraordinary commitment of the more than 2,000 authors and reviewers worldwide who contributed their knowledge, creativity, time, and enthusiasm to this process. We would like to express our gratitude to the members of the MA Assessment Panel, Coordinating Lead Authors, Lead Authors, Contributing Authors, Board of Review Editors, and Expert Reviewers who contributed to this process, and we wish to acknowledge the in-kind support of their institutions, which enabled their participation. (The list of reviewers is available at www.MAweb.org.) We also thank the members of the synthesis teams and the synthesis team co-chairs: Zafar Adeel, Carlos Corvalan, Rebecca D'Cruz, Nick Davidson, Anantha Kumar Duraiappah, C. Max Finlayson, Simon Hales, Jane Lubchenco, Anthony McMichael, Shahid Naeem, David Niemeijer, Steve Percy, Uriel Safriel, and Robin White.

We would like to thank the host organizations of the MA Technical Support Units—WorldFish Center (Malaysia); UNEP-World Conservation Monitoring Centre (United Kingdom); Institute of Economic Growth (India); National Institute of Public Health and the Environment (Netherlands); University of Pretoria (South Africa), U.N. Food and Agriculture Organization; World Resources Institute, Meridian Institute, and Center for Limnology of the University of Wisconsin (all in the United States); Scientific Committee on Problems of the Environment (France); and Interna-tional Maize and Wheat Improvement Center (Mexico)—for the support they provided to the process. The Scenarios Working Group was established as a joint project of the MA and the Scientific Committee on Problems of the Envi-ronment, and we thank SCOPE for the scientific input and oversight that it provided.

We thank the members of the MA Board (listed earlier) for the guidance and oversight they provided to this process and we also thank the current and previous Board Alternates: Ivar Baste, Jeroen Bordewijk, David Cooper, Carlos Corvalan, Nick Davidson, Lyle Glowka, Guo Risheng, Ju Hongbo, Ju Jin, Kagumaho (Bob) Kakuyo, Melinda Kimble, Kanta Kumari, Stephen Lonergan, Charles Ian McNeill, Joseph Kalemani Mulongoy, Ndegwa Ndiang'ui, and Mohamed Maged Younes. The contributions of past members of the MA Board were instrumental in shaping the MA focus and process and these individuals include Philbert Brown, Gisbert Glaser, He Changchui, Richard Helmer, Yolanda Kakabadse, Yoriko Kawaguchi, Ann Kern, Roberto Lenton, Corinne Lepage, Hubert Markl, Arnulf Müller-Helbrecht, Alfred Oteng-Yeboah, Seema Paul, Susan Pineda Mercado, Jan Plesnik, Peter Raven, Cristián Samper,

Ola Smith, Dennis Tirpak, Alvaro Umaña, and Meryl Williams. We wish to also thank the members of the Exploratory Steering Committee that designed the MA project in 1999–2000. This group included a number of the current and past Board members, as well as Edward Ayensu, Daniel Claasen, Mark Collins, Andrew Dearing, Louise Fresco, Madhav Gadgil, Habiba Gitay, Zuzana Guziova, Calestous Juma, John Krebs, Jane Lubchenco, Jeffrey McNeely, Ndegwa Ndiang'ui, Janos Pasztor, Prabhu L. Pingali, Per Pinstrup-Andersen, and José Sarukhán. And we would like to acknowledge the support and guidance provided by the secretariats and the scientific and technical bodies of the Convention on Biological Diversity, the Ramsar Convention on Wetlands, the Convention to Combat Desertification, and the Convention on Migratory Species, which have helped to define the focus of the MA and of this report. We are grateful to two members of the Board of Review Editors, Gordon Orians and Richard Norgaard, who played a particularly important role during the review and revision of this synthesis report. And, we would like to thank Ian Noble and Mingsarn Kaosa-ard for their contributions as members of the Assessment Panel during 2002.

We thank the interns and volunteers who worked with the MA Secretariat, part-time members of the Secretariat staff, the administrative staff of the host organizations, and colleagues in other organizations who were instrumental in facilitating the process: Isabelle Alegre, Adlai Amor, Hyacinth Billings, Cecilia Blasco, Delmar Blasco, Herbert Caudill, Lina Cimarrusti, Emily Cooper, Dalène du Plessis, Keisha-Maria Garcia, Habiba Gitay, Helen Gray, Sherry Heileman, Norbert Henninger, Tim Hirsch, Toshie Honda, Francisco Ingouville, Humphrey Kagunda, Brygida Kubiak, Nicholas Lapham, Liz Levitt, Christian Marx, Stephanie Moore, John Mukoza, Arivudai Nambi, Laurie Neville, Rosemarie Philips, Veronique Plocq Fichelet, Maggie Powell, Janet Ranganathan, Carolina Katz Reid, Liana Reilly, Carol Rosen, Mariana Sanchez Abregu, Anne Schram, Jean Sedgwick, Tang Siang Nee, Darrell Taylor, Tutti Tischler, Daniel Tunstall, Woody Turner, Mark Valentine, Elsie Vélez-Whited, Elizabeth Wilson, and Mark Zimsky. Special thanks are due to Linda Starke, who skillfully edited this report, and to Philippe Rekacewicz and Emmanuelle Bournay of UNEP/GRID-Arendal, who prepared the Figures.

We also want to acknowledge the support of a large number of nongovernmental organizations and networks around the world that have assisted in outreach efforts: Alexandria University, Argentine Business Council for Sustainable Development, Asociación Ixa Ca Vaá (Costa Rica), Arab Media Forum for Environment and Development, Brazilian Business Council on Sustainable Development, Charles University (Czech Republic), Chinese Academy of Sciences, European Environmental Agency, European Union of Science Journalists' Associations, EIS-Africa (Burkina Faso), Forest Institute of the State of São Paulo, Foro Ecológico (Peru), Fridtjof Nansen Institute (Norway), Fundación Natura (Ecuador), Global Development Learning Network, Indonesian Biodiversity Foundation, Institute for Biodiversity Conservation and Research–Academy of Sciences of Bolivia, International Alliance of Indigenous Peoples of the Tropical Forests, IUCN office in Uzbekistan, IUCN Regional Offices for West Africa and South America, Permanent Inter-States Committee for Drought Control in the Sahel, Peruvian Society of Environmental Law, Probioandes (Peru), Professional Council of Environmental Analysts of Argentina, Regional Center AGRHYMET (Niger), Regional Environmental Centre for Central Asia, Resources and Research for Sustainable Development (Chile), Royal Society (United Kingdom), Stockholm University, Suez Canal University, Terra Nuova (Nicaragua), The Nature Conservancy (United States), United Nations University, University of Chile, University of the Philippines, World Assembly of Youth, World Business Council for Sustainable Development, WWF-Brazil, WWF-Italy, and WWF-US.

We are extremely grateful to the donors that provided major financial support for the MA and the MA Sub-global Assessments: Global Environment Facility; United Nations Foundation; The David and Lucile Packard Foundation; The World Bank; Consultative Group on International Agricultural Research; United Nations Environment Programme; Government of China; Ministry of Foreign Affairs of the Government of Norway; Kingdom of Saudi Arabia;

and the Swedish International Biodiversity Programme. We also thank other organizations that provided financial support: Asia Pacific Network for Global Change Research; Association of Caribbean States; British High Commission, Trinidad and Tobago; Caixa Geral de Depósitos, Portugal; Canadian International Development Agency; Christensen Fund; Cropper Foundation, Environmental Management Authority of Trinidad and Tobago; Ford Foundation; Government of India; International Council for Science; International Development Research Centre; Island Resources Foundation; Japan Ministry of Environment; Laguna Lake Development Authority; Philippine Department of Environment and Natural Resources; Rockefeller Foundation; U.N. Educational, Scientific and Cultural Organization; UNEP Division of Early Warning and Assessment; United Kingdom Department for Environment, Food and Rural Affairs; United States National Aeronautic and Space Administration; and Universidade de Coimbra, Portugal. Generous in-kind support has been provided by many other institutions (a full list is available at www.MAweb.org). The work to establish and design the MA was supported by grants from The Avina Group, The David and Lucile Packard Foundation, Global Environment Facility, Directorate for Nature Management of Norway, Swedish International Development Cooperation Authority, Summit Foundation, UNDP, UNEP, United Nations Foundation, United States Agency for International Development, Wallace Global Fund, and The World Bank.

We give special thanks for the extraordinary contributions of the coordinators and full-time staff of the MA Secretariat: Neville Ash, Elena Bennett, Chan Wai Leng, John Ehrmann, Lori Han, Christine Jalleh, Nicole Khi, Pushpam Kumar, Marcus Lee, Belinda Lim, Nicolas Lucas, Mampiti Matete, Tasha Merican, Meenakshi Rathore, Ciara Raudsepp-Hearne, Henk Simons, Sara Suriani, Jillian Thonell, Valerie Thompson, and Monika Zurek.

Finally, we would particularly like to thank Angela Cropper and Harold Mooney, the co-chairs of the MA Assessment Panel, and José Sarukhán and Anne Whyte, the co-chairs of the MA Review Board, for their skillful leadership of the assessment and review processes, and Walter Reid, the MA Director for his pivotal role in establishing the assessment, his leadership, and his outstanding contributions to the process.

DR. ROBERT T. WATSON
MA Board Co-chair
Chief Scientist
The World Bank

DR. A.H. ZAKRI
MA Board Co-chair
Director, Institute for Advanced Studies
United Nations University

PREFACE

The Millennium Ecosystem Assessment was carried out between 2001 and 2005 to assess the consequences of ecosystem change for human well-being and to establish the scientific basis for actions needed to enhance the conservation and sustainable use of ecosystems and their contributions to human well-being. The MA responds to government requests for information received through four international conventions—the Convention on Biological Diversity, the United Nations Convention to Combat Desertification, the Ramsar Convention on Wetlands, and the Convention on Migratory Species—and is designed to also meet needs of other stakeholders, including the business community, the health sector, nongovernmental organizations, and indigenous peoples. The sub-global assessments also aimed to meet the needs of users in the regions where they were undertaken.

The assessment focuses on the linkages between ecosystems and human well-being and, in particular, on "ecosystem services." An ecosystem is a dynamic complex of plant, animal, and microorganism communities and the nonliving environment interacting as a functional unit. The MA deals with the full range of ecosystems—from those relatively undisturbed, such as natural forests, to landscapes with mixed patterns of human use, to ecosystems intensively managed and modified by humans, such as agricultural land and urban areas. Ecosystem services are the benefits people obtain from ecosystems. These include *provisioning services* such as food, water, timber, and fiber; *regulating services* that affect climate, floods, disease, wastes, and water quality; *cultural services* that provide recreational, aesthetic, and spiritual benefits; and *supporting services* such as soil formation, photosynthesis, and nutrient cycling. (See Figure A.) The human species, while buffered against environmental changes by culture and technology, is fundamentally dependent on the flow of ecosystem services.

The MA examines how changes in ecosystem services influence human well-being. Human well-being is assumed to have multiple constituents, including the *basic material for a good life*, such as secure and adequate livelihoods, enough food at all times, shelter, clothing, and access to goods; *health*, including feeling well and having a healthy physical environment, such as clean air and access to clean water; *good social relations*, including social cohesion, mutual respect, and the ability to help others and provide for children; *security*, including secure access to natural and other resources, personal safety, and security from natural and human-made disasters; and *freedom of choice and action*, including the opportunity to achieve what an individual values doing and being. Freedom of choice and action is influenced by other constituents of well-being (as well as by other factors, notably education) and is also a precondition for achieving other components of well-being, particularly with respect to equity and fairness.

The conceptual framework for the MA posits that people are integral parts of ecosystems and that a dynamic interaction exists between them and other parts of ecosystems, with the changing human condition driving, both directly and indirectly, changes in ecosystems and thereby causing changes in human well-being. (See Figure B.) At the same time, social, economic, and cultural factors unrelated to ecosystems alter the human condition, and many natural forces influence ecosystems. Although the MA emphasizes the linkages between ecosystems and human well-being, it recognizes that the actions people take that influence ecosystems result not just from concern about human well-being but also from considerations of the intrinsic value of species and ecosystems. Intrinsic value is the value of something in and for itself, irrespective of its utility for someone else.

The Millennium Ecosystem Assessment synthesizes information from the scientific literature and relevant peer-reviewed datasets and models. It incorporates knowledge held by the private sector, practitioners, local communities, and indigenous peoples. The MA did not aim to generate new primary knowledge, but instead sought to add value to existing information by collating, evaluating, summarizing, interpreting, and communicating it in a useful form. Assessments like this one apply the judgment of experts to existing knowledge to provide scientifically credible answers to policy-relevant questions. The focus on policy-relevant questions and the explicit use of expert judgment distinguish this type of assessment from a scientific review.

This Figure depicts the strength of linkages between categories of ecosystem services and components of human well-being that are commonly encountered, and includes indications of the extent to which it is possible for socioeconomic factors to mediate the linkage. (For example, if it is possible to purchase a substitute for a degraded ecosystem service, then there is a high potential for mediation.) The strength of the linkages and the potential for mediation differ in different ecosystems and regions. In addition to the influence of ecosystem services on human well-being depicted here, other factors—including other environmental factors as well as economic, social, technological, and cultural factors—influence human well-being, and ecosystems are in turn affected by changes in human well-being. (See Figure B.)

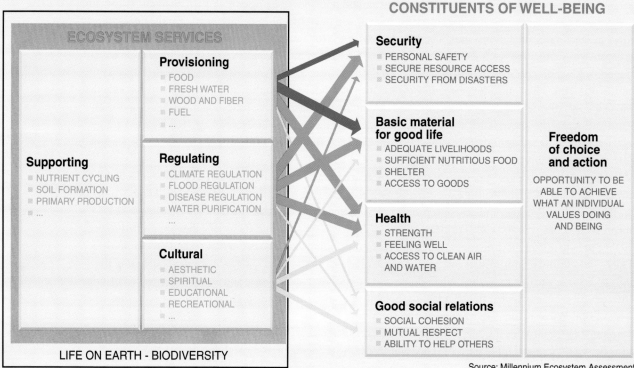

Source: Millennium Ecosystem Assessment

ARROW'S COLOR
Potential for mediation by socioeconomic factors

    Low

    Medium

    High

ARROW'S WIDTH
Intensity of linkages between ecosystem services and human well-being

    Weak

    Medium

    Strong

Changes in drivers that indirectly affect biodiversity, such as population, technology, and lifestyle (upper right corner of Figure), can lead to changes in drivers directly affecting biodiversity, such as the catch of fish or the application of fertilizers (lower right corner). These result in changes to ecosystems and the services they provide (lower left corner), thereby affecting human well-being. These interactions can take place at more than one scale and can cross scales. For example, an international demand for timber may lead to a regional loss of forest cover, which increases flood magnitude along a local stretch of a river. Similarly, the interactions can take place across different time scales. Different strategies and interventions can be applied at many points in this framework to enhance human well-being and conserve ecosystems.

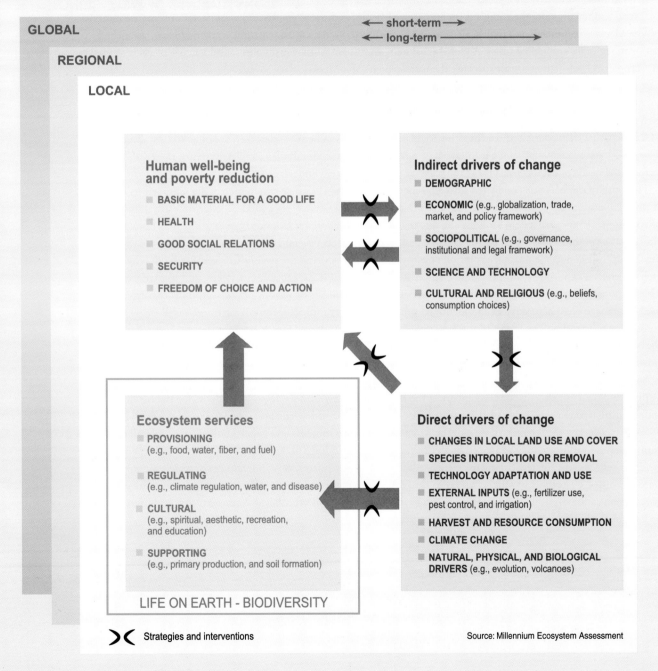

Source: Millennium Ecosystem Assessment

Five overarching questions, along with more detailed lists of user needs developed through discussions with stakeholders or provided by governments through international conventions, guided the issues that were assessed:

- What are the current condition and trends of ecosystems, ecosystem services, and human well-being?
- What are plausible future changes in ecosystems and their ecosystem services and the consequent changes in human well-being?
- What can be done to enhance well-being and conserve ecosystems? What are the strengths and weaknesses of response options that can be considered to realize or avoid specific futures?
- What are the key uncertainties that hinder effective decision-making concerning ecosystems?
- What tools and methodologies developed and used in the MA can strengthen capacity to assess ecosystems, the services they provide, their impacts on human well-being, and the strengths and weaknesses of response options?

The MA was conducted as a multiscale assessment, with interlinked assessments undertaken at local, watershed, national, regional, and global scales. A global ecosystem assessment cannot easily meet all the needs of decision-makers at national and sub-national scales because the management of any particular ecosystem must be tailored to the particular characteristics of that ecosystem and to the demands placed on it. However, an assessment focused only on a particular ecosystem or particular nation is insufficient because some processes are global and because local goods, services, matter, and energy are often transferred across regions. Each of the component assessments was guided by the MA conceptual framework and benefited from the presence of assessments undertaken at larger and smaller scales. The sub-global assessments were not intended to serve as representative samples of all ecosystems; rather, they were to meet the needs of decision-makers at the scales at which they were undertaken.

The work of the MA was conducted through four working groups, each of which prepared a report of its findings. At the global scale, the Condition and Trends Working Group assessed the state of knowledge on ecosystems, drivers of ecosystem change, ecosystem services, and associated human well-being around the year 2000. The assessment aimed to be comprehensive with regard to ecosystem services, but its coverage is not exhaustive. The Scenarios Working Group considered the possible evolution of ecosystem services during the twenty-first century by developing four global scenarios exploring plausible future changes in drivers, ecosystems, ecosystem services, and human well-being. The Responses Working Group examined the strengths and weaknesses of various response options that have been used to manage ecosystem services and identified promising opportunities for improving human well-being while conserving ecosystems. The report of the Sub-global Assessments Working Group contains lessons learned from the MA sub-global assessments. The first product of the MA—*Ecosystems and Human Well-being: A Framework for Assessment,* published in 2003—outlined the focus, conceptual basis, and methods used in the MA.

Approximately 1,360 experts from 95 countries were involved as authors of the assessment reports, as participants in the sub-global assessments, or as members of the Board of Review Editors. (See Appendix C for the list of coordinating lead authors, sub-global assessment coordinators, and review editors.) The latter group, which involved 80 experts, oversaw the scientific review of the MA reports by governments and experts and ensured that all review comments were appropriately addressed by the authors. All MA findings underwent two rounds of expert and governmental review. Review comments were received from approximately 850 individuals (of which roughly 250 were submitted by authors of other chapters in the MA), although in a number of cases (particularly in the case of governments and MA-affiliated scientific organizations), people submitted collated comments that had been prepared by a number of reviewers in their governments or institutions.

The MA was guided by a Board that included representatives of five international conventions, five U.N. agencies, international scientific organizations, governments, and leaders from the private sector, nongovernmental organizations, and indigenous groups. A 15-member Assessment Panel of leading social and natural scientists oversaw the technical work of the assessment, supported by a secretariat with offices in Europe, North America, South America, Asia, and Africa and coordinated by the United Nations Environment Programme.

The MA is intended to be used:

- to identify priorities for action;
- as a benchmark for future assessments;
- as a framework and source of tools for assessment, planning, and management;
- to gain foresight concerning the consequences of decisions affecting ecosystems;
- to identify response options to achieve human development and sustainability goals;
- to help build individual and institutional capacity to undertake integrated ecosystem assessments and act on the findings; and
- to guide future research.

Because of the broad scope of the MA and the complexity of the interactions between social and natural systems, it proved to be difficult to provide definitive information for some of the issues addressed in the MA. Relatively few ecosystem services have been the focus of research and monitoring and, as a consequence, research findings and data are often inadequate for a detailed global assessment. Moreover, the data and information that are available are generally related to either the characteristics of the ecological system or the characteristics of the social system, not to the all-important interactions between these systems. Finally, the scientific and assessment tools and models available to undertake a cross-scale integrated assessment and to project future changes in ecosystem services are only now being developed. Despite these challenges, the MA was able to provide considerable information relevant to most of the focal questions. And by identifying gaps in data and information that prevent policy-relevant questions from being answered, the assessment can help to guide research and monitoring that may allow those questions to be answered in future assessments.

READER'S GUIDE

This report presents a synthesis and integration of the findings of the four MA Working Groups along with more detailed findings for selected ecosystem services concerning condition and trends and scenarios (see Appendix A) and response options (see Appendix B). Five additional synthesis reports were prepared for ease of use by specific audiences: CBD (biodiversity), UNCCD (desertification), Ramsar Convention (wetlands), business, and the health sector. Each MA sub-global assessment will also produce additional reports to meet the needs of its own audience. The full technical assessment reports of the four MA Working Groups will be published in mid-2005 by Island Press. All printed materials of the assessment, along with core data and a glossary of terminology used in the technical reports, will be available on the Internet at www.MAweb.org. Appendix D lists the acronyms and abbreviations used in this report and includes additional information on sources for some of the Figures. Throughout this report, dollar signs indicate U.S. dollars and tons mean metric tons.

References that appear in parentheses in the body of this synthesis report are to the underlying chapters in the full technical assessment reports of each Working Group. (A list of the assessment report chapters is provided in Appendix E.) To assist the reader, citations to the technical volumes generally specify sections of chapters or specific Boxes, Tables, or Figures, based on final drafts of the chapter. Some chapter subsection numbers may change during final copyediting, however, after this synthesis report has been printed. Bracketed references within the Summary for Decision-makers are to the key questions of this full synthesis report, where additional information on each topic can be found.

In this report, the following words have been used where appropriate to indicate judgmental estimates of certainty, based on the collective judgment of the authors, using the observational evidence, modeling results, and theory that they have examined: very certain (98% or greater probability), high certainty (85–98% probability), medium certainty (65–85% probability), low certainty (52–65% probability), and very uncertain (50–52% probability). In other instances, a qualitative scale to gauge the level of scientific understanding is used: well established, established but incomplete, competing explanations, and speculative. Each time these terms are used they appear in italics.

SUMMARY FOR DECISION-MAKERS

Everyone in the world depends completely on Earth's ecosystems and the services they provide, such as food, water, disease management, climate regulation, spiritual fulfillment, and aesthetic enjoyment. Over the past 50 years, humans have changed these ecosystems more rapidly and extensively than in any comparable period of time in human history, largely to meet rapidly growing demands for food, fresh water, timber, fiber, and fuel. This transformation of the planet has contributed to substantial net gains in human well-being and economic development. But not all regions and groups of people have benefited from this process—in fact, many have been harmed. Moreover, the full costs associated with these gains are only now becoming apparent.

Three major problems associated with our management of the world's ecosystems are already causing significant harm to some people, particularly the poor, and unless addressed will substantially diminish the long-term benefits we obtain from ecosystems:

■ First, approximately 60% (15 out of 24) of the ecosystem services examined during the Millennium Ecosystem Assessment are being degraded or used unsustainably, including fresh water, capture fisheries, air and water purification, and the regulation of regional and local climate, natural hazards, and pests. The full costs of the loss and degradation of these ecosystem services are difficult to measure, but the available evidence demonstrates that they are substantial and growing. Many ecosystem services have been degraded as a consequence of actions taken to increase the supply of other services, such as food. These trade-offs often shift the costs of degradation from one group of people to another or defer costs to future generations.

■ Second, there is *established but incomplete* evidence that changes being made in ecosystems are increasing the likelihood of nonlinear changes in ecosystems (including accelerating, abrupt, and potentially irreversible changes) that have important consequences for human well-being. Examples of such changes include disease emergence, abrupt alterations in water quality, the creation of "dead zones" in coastal waters, the collapse of fisheries, and shifts in regional climate.

Four Main Findings

■ Over the past 50 years, humans have changed ecosystems more rapidly and extensively than in any comparable period of time in human history, largely to meet rapidly growing demands for food, fresh water, timber, fiber, and fuel. This has resulted in a substantial and largely irreversible loss in the diversity of life on Earth.

■ The changes that have been made to ecosystems have contributed to substantial net gains in human well-being and economic development, but these gains have been achieved at growing costs in the form of the degradation of many ecosystem services, increased risks of nonlinear changes, and the exacerbation of poverty for some groups of people. These problems, unless addressed, will substantially diminish the benefits that future generations obtain from ecosystems.

■ The degradation of ecosystem services could grow significantly worse during the first half of this century and is a barrier to achieving the Millennium Development Goals.

■ The challenge of reversing the degradation of ecosystems while meeting increasing demands for their services can be partially met under some scenarios that the MA has considered, but these involve significant changes in policies, institutions, and practices that are not currently under way. Many options exist to conserve or enhance specific ecosystem services in ways that reduce negative trade-offs or that provide positive synergies with other ecosystem services.

■ Third, the harmful effects of the degradation of ecosystem services (the persistent decrease in the capacity of an ecosystem to deliver services) are being borne disproportionately by the poor, are contributing to growing inequities and disparities across groups of people, and are sometimes the principal factor causing poverty and social conflict. This is not to say that ecosystem changes such as increased food production have not also helped to lift many people out of poverty or hunger, but these changes have harmed other individuals and communities, and their plight has been largely overlooked. In all regions, and particularly in sub-Saharan Africa, the condition and management of ecosystem services is a dominant factor influencing prospects for reducing poverty.

The degradation of ecosystem services is already a significant barrier to achieving the Millennium Development Goals agreed to by the international community in September 2000 and the harmful consequences of this degradation could grow significantly worse in the next 50 years. The consumption of ecosystem services, which is unsustainable in many cases, will continue to grow as a consequence of a likely three- to sixfold increase in global GDP by 2050 even while global population growth is expected to slow and level off in mid-century. Most of the important direct drivers of ecosystem change are unlikely to diminish in the first half of the century and two drivers—climate change and excessive nutrient loading—will become more severe.

Already, many of the regions facing the greatest challenges in achieving the MDGs coincide with those facing significant problems of ecosystem degradation. Rural poor people, a primary target of the MDGs, tend to be most directly reliant on ecosystem services and most vulnerable to changes in those services. More generally, any progress achieved in addressing the MDGs of poverty and hunger eradication, improved health, and environmental sustainability is unlikely to be sustained if most of the ecosystem services on which humanity relies continue to be degraded. In contrast, the sound management of ecosystem services provides cost-effective opportunities for addressing multiple development goals in a synergistic manner.

There is no simple fix to these problems since they arise from the interaction of many recognized challenges, including climate change, biodiversity loss, and land degradation, each of which is complex to address in its own right. Past actions to slow or reverse the degradation of ecosystems have yielded significant benefits, but these improvements have generally not kept pace with growing pressures and demands. Nevertheless, there is tremendous scope for action to reduce the severity of these problems in the coming decades. Indeed, three of four detailed scenarios examined by the MA suggest that significant changes in policies, institutions, and practices can mitigate some but not all of the negative consequences of growing pressures on ecosystems. But the changes required are substantial and are not currently under way.

An effective set of responses to ensure the sustainable management of ecosystems requires substantial changes in institutions and governance, economic policies and incentives, social and behavior factors, technology, and knowledge. Actions such as the integration of ecosystem management goals in various sectors (such as agriculture, forestry, finance, trade, and health), increased transparency and accountability of government and private-sector performance in ecosystem management, elimination of perverse subsidies, greater use of economic instruments and market-based approaches, empowerment of groups dependent on ecosystem services or affected by their degradation, promotion of technologies enabling increased crop yields without harmful environmental impacts, ecosystem restoration, and the incorporation of nonmarket values of ecosystems and their services in management decisions all could substantially lessen the severity of these problems in the next several decades.

The remainder of this Summary for Decision-makers presents the four major findings of the Millennium Ecosystem Assessment on the problems to be addressed and the actions needed to enhance the conservation and sustainable use of ecosystems.

> **Finding #1:** *Over the past 50 years, humans have changed ecosystems more rapidly and extensively than in any comparable period of time in human history, largely to meet rapidly growing demands for food, fresh water, timber, fiber, and fuel. This has resulted in a substantial and largely irreversible loss in the diversity of life on Earth.*

The structure and functioning of the world's ecosystems changed more rapidly in the second half of the twentieth century than at any time in human history. [1]

■ More land was converted to cropland in the 30 years after 1950 than in the 150 years between 1700 and 1850. Cultivated systems (areas where at least 30% of the landscape is in croplands, shifting cultivation, confined livestock production, or freshwater aquaculture) now cover one quarter of Earth's terrestrial surface. (See Figure 1.) Areas of rapid change in forest land cover and land degradation are shown in Figure 2.

■ Approximately 20% of the world's coral reefs were lost and an additional 20% degraded in the last several decades of the twentieth century, and approximately 35% of mangrove area was lost during this time (in countries for which sufficient data exist, which encompass about half of the area of mangroves).

■ The amount of water impounded behind dams quadrupled since 1960, and three to six times as much water is held in reservoirs as in natural rivers. Water withdrawals from rivers and lakes doubled since 1960; most water use (70% worldwide) is for agriculture.

■ Since 1960, flows of reactive (biologically available) nitrogen in terrestrial ecosystems have doubled, and flows of phosphorus have tripled. More than half of all the synthetic nitrogen fertilizer, which was first manufactured in 1913, ever used on the planet has been used since 1985.

Figure 1. EXTENT OF CULTIVATED SYSTEMS, 2000. Cultivated systems cover 24% of the terrestrial surface.

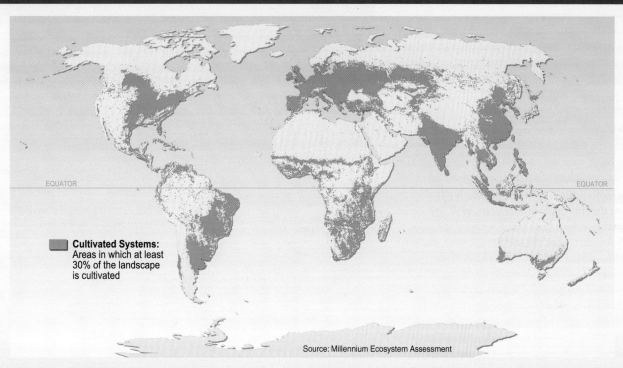

Cultivated Systems: Areas in which at least 30% of the landscape is cultivated

Source: Millennium Ecosystem Assessment

Figure 2. LOCATIONS REPORTED BY VARIOUS STUDIES AS UNDERGOING HIGH RATES OF LAND COVER CHANGE IN THE PAST FEW DECADES (C.SDM)

In the case of forest cover change, the studies refer to the period 1980–2000 and are based on national statistics, remote sensing, and to a limited degree expert opinion. In the case of land cover change resulting from degradation in drylands (desertification), the period is unspecified but inferred to be within the last half-century, and the major study was entirely based on expert opinion, with associated *low certainty*. Change in cultivated area is not shown. Note that areas showing little current change are often locations that have already undergone major historical change (see Figure 1).

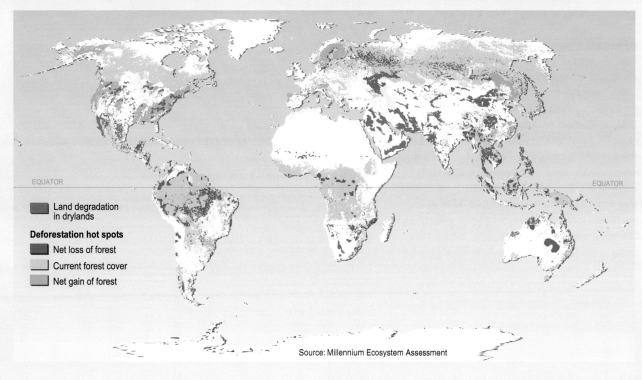

Land degradation in drylands

Deforestation hot spots

Net loss of forest

Current forest cover

Net gain of forest

Source: Millennium Ecosystem Assessment

■ Since 1750, the atmospheric concentration of carbon dioxide has increased by about 32% (from about 280 to 376 parts per million in 2003), primarily due to the combustion of fossil fuels and land use changes. Approximately 60% of that increase (60 parts per million) has taken place since 1959.

Humans are fundamentally, and to a significant extent irreversibly, changing the diversity of life on Earth, and most of these changes represent a loss of biodiversity. [1]

■ More than two thirds of the area of 2 of the world's 14 major terrestrial biomes and more than half of the area of 4 other biomes had been converted by 1990, primarily to agriculture. (See Figure 3.)

■ Across a range of taxonomic groups, either the population size or range or both of the majority of species is currently declining.

■ The distribution of species on Earth is becoming more homogenous; in other words, the set of species in any one region of the world is becoming more similar to the set in other regions primarily as a result of introductions of species, both intentionally and inadvertently in association with increased travel and shipping.

■ The number of species on the planet is declining. Over the past few hundred years, humans have increased the species extinction rate by as much as 1,000 times over background rates typical over the planet's history (*medium certainty*). (See Figure 4.) Some 10–30% of mammal, bird, and amphibian species are currently threatened with extinction (*medium to high certainty*). Freshwater ecosystems tend to have the highest proportion of species threatened with extinction.

■ Genetic diversity has declined globally, particularly among cultivated species.

Most changes to ecosystems have been made to meet a dramatic growth in the demand for food, water, timber, fiber, and fuel. [2] Some ecosystem changes have been the inadvertent result of activities unrelated to the use of ecosystem services, such as the construction of roads, ports, and cities and the discharge of pollutants. But most ecosystem changes were the direct or indirect result of changes made to meet growing demands for ecosystem services, and in particular growing demands for food, water, timber, fiber, and fuel (fuelwood and hydropower).

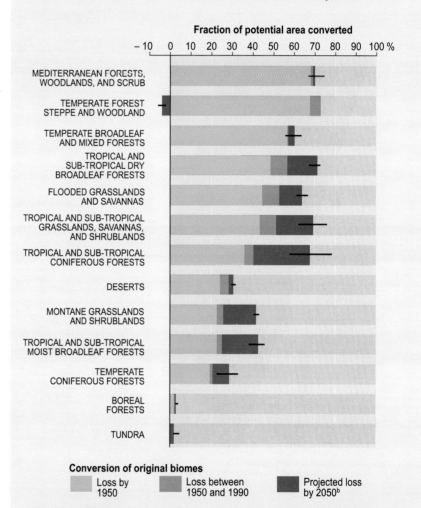

Figure 3. Conversion of Terrestrial Biomes[a]
(Adapted from C4, S10)

It is not possible to estimate accurately the extent of different biomes prior to significant human impact, but it is possible to determine the "potential" area of biomes based on soil and climatic conditions. This Figure shows how much of that potential area is estimated to have been converted by 1950 (*medium certainty*), how much was converted between 1950 and 1990 (*medium certainty*), and how much would be converted under the four MA scenarios (*low certainty*) between 1990 and 2050. Mangroves are not included here because the area was too small to be accurately assessed. Most of the conversion of these biomes is to cultivated systems.

Conversion of original biomes

- Loss by 1950
- Loss between 1950 and 1990
- Projected loss by 2050[b]

[a] A biome is the largest unit of ecological classification that is convenient to recognize below the entire globe, such as temperate broadleaf forests or montane grasslands. A biome is a widely used ecological categorization, and because considerable ecological data have been reported and modeling undertaken using this categorization, some information in this assessment can only be reported based on biomes. Whenever possible, however, the MA reports information using 10 socioecological systems, such as forest, cultivated, coastal, and marine, because these correspond to the regions of responsibility of different government ministries and because they are the categories used within the Convention on Biological Diversity.

[b] According to the four MA scenarios. For 2050 projections, the average value of the projections under the four scenarios is plotted and the error bars (black lines) represent the range of values from the different scenarios.

Source: Millennium Ecosystem Assessment

Between 1960 and 2000, the demand for ecosystem services grew significantly as world population doubled to 6 billion people and the global economy increased more than sixfold. To meet this demand, food production increased by roughly two-and-a-half times, water use doubled, wood harvests for pulp and paper production tripled, installed hydropower capacity doubled, and timber production increased by more than half.

The growing demand for these ecosystem services was met both by consuming an increasing fraction of the available supply (for example, diverting more water for irrigation or capturing more fish from the sea) and by raising the production of some services, such as crops and livestock. The latter has been accomplished through the use of new technologies (such as new crop varieties, fertilization, and irrigation) as well as through increasing the area managed for the services in the case of crop and livestock production and aquaculture.

Finding #2: *The changes that have been made to ecosystems have contributed to substantial net gains in human well-being and economic development, but these gains have been achieved at growing costs in the form of the degradation of many ecosystem services, increased risks of nonlinear changes, and the exacerbation of poverty for some groups of people. These problems, unless addressed, will substantially diminish the benefits that future generations obtain from ecosystems.*

In the aggregate, and for most countries, changes made to the world's ecosystems in recent decades have provided substantial benefits for human well-being and national development. [3] Many of the most significant changes to ecosystems have been essential to meet growing needs for food and water; these

---

Figure 4. SPECIES EXTINCTION RATES (Adapted from C4 Fig 4.22)

"Distant past" refers to average extinction rates as estimated from the fossil record. "Recent past" refers to extinction rates calculated from known extinctions of species (lower estimate) or known extinctions plus "possibly extinct" species (upper bound). A species is considered to be "possibly extinct" if it is believed by experts to be extinct but extensive surveys have not yet been undertaken to confirm its disappearance. "Future" extinctions are model-derived estimates using a variety of techniques, including species-area models, rates at which species are shifting to increasingly more threatened categories, extinction probabilities associated with the IUCN categories of threat, impacts of projected habitat loss on species currently threatened with habitat loss, and correlation of species loss with energy consumption. The time frame and species groups involved differ among the "future" estimates, but in general refer to either future loss of species based on the level of threat that exists

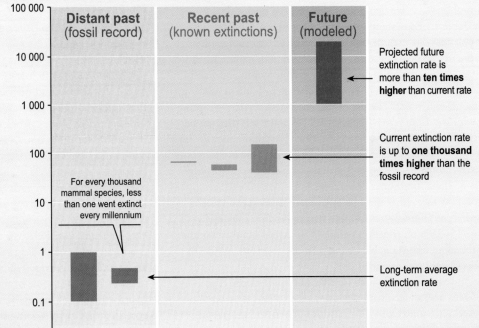

Source: Millennium Ecosystem Assessment

today or current and future loss of species as a result of habitat changes taking place over the period of roughly 1970 to 2050. Estimates based on the fossil record are *low certainty*; lower-bound estimates for known extinctions are *high certainty* and upper-bound estimates are *medium certainty*; lower-bound estimates for modeled extinctions are *low certainty* and upper-bound estimates are *speculative*. The rate of known extinctions of species in the past century is roughly 50–500 times greater than the extinction rate calculated from the fossil record of 0.1–1 extinctions per 1,000 species per 1,000 years. The rate is up to 1,000 times higher than the background extinction rates if possibly extinct species are included.

changes have helped reduce the proportion of malnourished people and improved human health. Agriculture, including fisheries and forestry, has been the mainstay of strategies for the development of countries for centuries, providing revenues that have enabled investments in industrialization and poverty alleviation. Although the value of food production in 2000 was only about 3% of gross world product, the agricultural labor force accounts for approximately 22% of the world's population, half the world's total labor force, and 24% of GDP in countries with per capita incomes of less than $765 (the low-income developing countries, as defined by the World Bank).

These gains have been achieved, however, at growing costs in the form of the degradation of many ecosystem services, increased risks of nonlinear changes in ecosystems, the exacerbation of poverty for some people, and growing inequities and disparities across groups of people.

Degradation and Unsustainable Use of Ecosystem Services

Approximately 60% (15 out of 24) of the ecosystem services evaluated in this assessment (including 70% of regulating and cultural services) are being degraded or used unsustainably. [2] (See Table 1.) Ecosystem services that have been degraded over the past 50 years include capture fisheries, water supply, waste treatment and detoxification, water purification, natural hazard protection, regulation of air quality, regulation of regional and local climate, regulation of erosion, spiritual fulfillment, and aesthetic enjoyment. The use of two ecosystem services—capture fisheries and fresh water—is now well beyond levels that can be sustained even at current demands, much less future ones. At least one quarter of important commercial fish stocks are overharvested (*high certainty*). (See Figures 5, 6, and 7.) From 5% to possibly 25% of global freshwater use exceeds long-term accessible supplies and is now met either through engineered water transfers or overdraft of groundwater supplies (*low to medium certainty*). Some 15–35% of irrigation withdrawals exceed supply rates and are therefore unsustainable (*low to medium certainty*). While 15 services have been degraded, only 4 have been enhanced in the past 50 years, three of which involve food production: crops, livestock, and aquaculture. Terrestrial ecosystems were on average a net source of CO_2 emissions during the nineteenth and early twentieth centuries, but became a net sink around the middle of the last century, and thus in the last 50 years the role of ecosystems in regulating global climate through carbon sequestration has also been enhanced.

Actions to increase one ecosystem service often cause the degradation of other services. [2, 6] For example, because actions to increase food production typically involve increased use of water and fertilizers or expansion of the area of cultivated land,

these same actions often degrade other ecosystem services, including reducing the availability of water for other uses, degrading water quality, reducing biodiversity, and decreasing forest cover (which in turn may lead to the loss of forest products and the release of greenhouse gasses). Similarly, the conversion of forest to agriculture can significantly change the frequency and magnitude of floods, although the nature of this impact depends on the characteristics of the local ecosystem and the type of land cover change.

The degradation of ecosystem services often causes significant harm to human well-being. [3, 6] The information available to assess the consequences of changes in ecosystem services for human well-being is relatively limited. Many ecosystem services have not been monitored, and it is also difficult to estimate the influence of changes in ecosystem services relative to other social, cultural, and economic factors that also affect human well-being. Nevertheless, the following types of evidence demonstrate that the harmful effects of the degradation of ecosystem services on livelihoods, health, and local and national economies are substantial.

■ *Most resource management decisions are most strongly influenced by ecosystem services entering markets; as a result, the nonmarketed benefits are often lost or degraded. These nonmarketed benefits are often high and sometimes more valuable than the marketed ones.* For example, one of the most comprehensive studies to date, which examined the marketed and nonmarketed economic values associated with forests in eight Mediterranean countries, found that timber and fuelwood generally accounted for less than a third of total economic value of forests in each country. (See Figure 8.) Values associated with non-wood forest products, recreation, hunting, watershed protection, carbon sequestration, and passive use (values independent of direct uses) accounted for between 25% and 96% of the total economic value of the forests.

■ *The total economic value associated with managing ecosystems more sustainably is often higher than the value associated with the conversion of the ecosystem through farming, clear-cut logging, or other intensive uses.* Relatively few studies have compared the total economic value (including values of both marketed and nonmarketed ecosystem services) of ecosystems under alternate management regimes, but some of the studies that do exist have found that the benefit of managing the ecosystem more sustainably exceeded that of converting the ecosystem. (See Figure 9.)

■ *The economic and public health costs associated with damage to ecosystem services can be substantial.*

- The early 1990s collapse of the Newfoundland cod fishery due to overfishing resulted in the loss of tens of thousands of jobs and cost at least $2 billion in income support and retraining.
- In 1996, the cost of U.K. agriculture resulting from the damage that agricultural practices cause to water (pollution and eutrophication, a process whereby excessive plant growth depletes oxygen in the water), air (emissions of greenhouse gases), soil (off-site erosion damage, emissions

Status indicates whether the condition of the service globally has been enhanced (if the productive capacity of the service has been increased, for example) or degraded in the recent past. Definitions of "enhanced" and "degraded" are provided in the note below. A fourth category, supporting services, is not included here as they are not used directly by people.

Service	Sub-category	Status	Notes
Provisioning Services			
Food	crops	▲	substantial production increase
	livestock	▲	substantial production increase
	capture fisheries	▼	declining production due to overharvest
	aquaculture	▲	substantial production increase
	wild foods	▼	declining production
Fiber	timber	+/–	forest loss in some regions, growth in others
	cotton, hemp, silk	+/–	declining production of some fibers, growth in others
	wood fuel	▼	declining production
Genetic resources		▼	lost through extinction and crop genetic resource loss
Biochemicals, natural medicines, pharmaceuticals		▼	lost through extinction, overharvest
Fresh water		▼	unsustainable use for drinking, industry, and irrigation; amount of hydro energy unchanged, but dams increase ability to use that energy
Regulating Services			
Air quality regulation		▼	decline in ability of atmosphere to cleanse itself
Climate regulation	global	▲	net source of carbon sequestration since mid-century
	regional and local	▼	preponderance of negative impacts
Water regulation		+/–	varies depending on ecosystem change and location
Erosion regulation		▼	increased soil degradation
Water purification and waste treatment		▼	declining water quality
Disease regulation		+/–	varies depending on ecosystem change
Pest regulation		▼	natural control degraded through pesticide use
Pollination		▼[a]	apparent global decline in abundance of pollinators
Natural hazard regulation		▼	loss of natural buffers (wetlands, mangroves)
Cultural Services			
Spiritual and religious values		▼	rapid decline in sacred groves and species
Aesthetic values		▼	decline in quantity and quality of natural lands
Recreation and ecotourism		+/–	more areas accessible but many degraded

Note: For provisioning services, we define enhancement to mean increased production of the service through changes in area over which the service is provided (e.g., spread of agriculture) or increased production per unit area. We judge the production to be degraded if the current use exceeds sustainable levels. For regulating and supporting services, enhancement refers to a change in the service that leads to greater benefits for people (e.g., the service of disease regulation could be improved by eradication of a vector known to transmit a disease to people). Degradation of regulating and supporting services means a reduction in the benefits obtained from the service, either through a change in the service (e.g., mangrove loss reducing the storm protection benefits of an ecosystem) or through human pressures on the service exceeding its limits (e.g., excessive pollution exceeding the capability of ecosystems to maintain water quality). For cultural services, enhancement refers to a change in the ecosystem features that increase the cultural (recreational, aesthetic, spiritual, etc.) benefits provided by the ecosystem.

[a] Indicates *low to medium certainty*. All other trends are *medium to high certainty*.

Figure 5. ESTIMATED GLOBAL MARINE FISH CATCH, 1950–2001 (C18 Fig 18.3)

In this Figure, the catch reported by governments is in some cases adjusted to correct for likely errors in data.

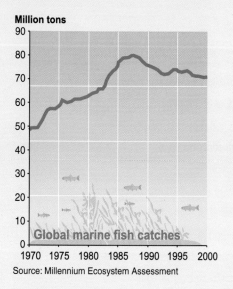

Source: Millennium Ecosystem Assessment

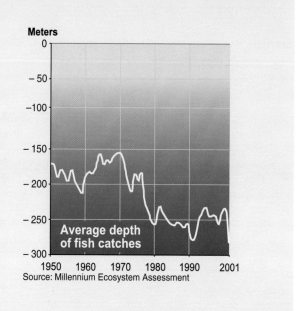

Source: Millennium Ecosystem Assessment

Figure 6. DECLINE IN TROPHIC LEVEL OF FISHERIES CATCH SINCE 1950 (C18)

A trophic level of an organism is its position in a food chain. Levels are numbered according to how far particular organisms are along the chain from the primary producers at level 1, to herbivores (level 2), to predators (level 3), to carnivores or top carnivores (level 4 or 5). Fish at higher trophic levels are typically of higher economic value. The decline in the trophic level harvested is largely a result of the overharvest of fish at higher trophic levels.

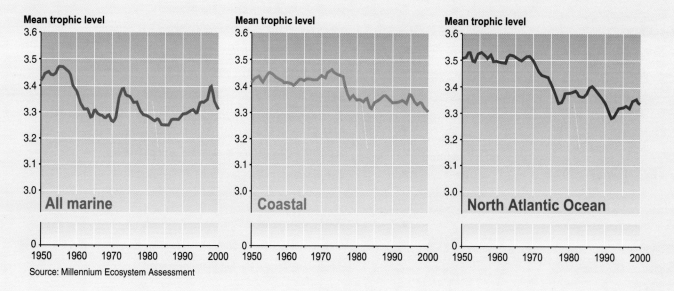

Source: Millennium Ecosystem Assessment

of greenhouse gases), and biodiversity was $2.6 billion, or 9% of average yearly gross farm receipts for the 1990s. Similarly, the damage costs of freshwater eutrophication alone in England and Wales (involving factors including reduced value of waterfront dwellings, water treatment costs, reduced recreational value of water bodies, and tourism losses) was estimated to be $105–160 million per year in the 1990s, with an additional $77 million a year being spent to address those damages.

- The incidence of diseases of marine organisms and the emergence of new pathogens is increasing, and some of these, such as ciguatera, harm human health. Episodes of harmful (including toxic) algal blooms in coastal waters are increasing in frequency and intensity, harming other marine resources such as fisheries as well as human health. In a particularly severe outbreak in Italy in 1989, harmful algal blooms cost the coastal aquaculture industry $10 million and the Italian tourism industry $11.4 million.
- The frequency and impact of floods and fires has increased significantly in the past 50 years, in part due to ecosystem changes. Examples are the increased susceptibility of coastal populations to tropical storms when mangrove forests are cleared and the increase in downstream flooding that followed land use changes in the upper Yangtze River. Annual economic losses from extreme events increased tenfold from the 1950s to approximately $70 billion in 2003, of which natural catastrophes (floods, fires, storms, drought, earthquakes) accounted for 84% of insured losses.

■ *The impact of the loss of cultural services is particularly difficult to measure, but it is especially important for many people.* Human cultures, knowledge systems, religions, and social interactions have been strongly influenced by ecosystems. A number of the MA sub-global assessments found that spiritual and cultural values of ecosystems were as important as other services for many local communities, both in developing countries (the importance of sacred groves of forest in India, for example) and industrial ones (the importance of urban parks, for instance).

The degradation of ecosystem services represents loss of a capital asset. [3] Both renewable resources such as ecosystem services and nonrenewable resources such as mineral deposits, some soil nutrients, and fossil fuels are capital assets. Yet traditional national accounts do not include measures of resource depletion or of the degradation of these resources. As a result, a country could cut its forests and deplete its fisheries, and this would show only as a positive gain in GDP (a measure of current economic well-being) without registering the corresponding decline in assets (wealth) that is the more appropriate measure of future economic well-being. Moreover, many ecosystem services (such as fresh water in aquifers and the use of the atmosphere as a sink for pollutants) are available freely to those who use them, and so again their degradation is not reflected in standard economic measures.

When estimates of the economic losses associated with the depletion of natural assets are factored into measurements of the total wealth of nations, they significantly change the balance sheet of countries with economies significantly dependent on natural resources. For example, countries such as Ecuador, Ethiopia, Kazakhstan, Democratic Republic of Congo, Trinidad and Tobago, Uzbekistan, and Venezuela that had positive growth in net savings in 2001, reflecting a growth in the net wealth of the country, actually experienced a loss in net savings when depletion of natural resources (energy and forests) and estimated damages from carbon emissions (associated with contributions to climate change) were factored into the accounts.

Figure 8. Annual Flow of Benefits from Forests in Selected Countries (Adapted from C5 Box 5.2)

In most countries, the marketed values of ecosystems associated with timber and fuelwood production are less than one third of the total economic value, including nonmarketed values such as carbon sequestration, watershed protection, and recreation.

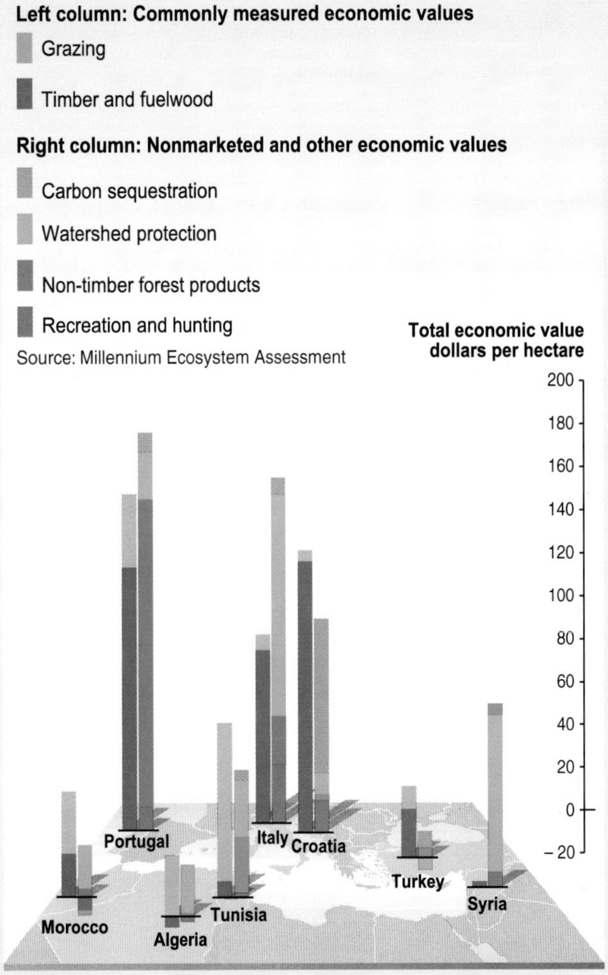
Left column: Commonly measured economic values
Grazing
Timber and fuelwood
Right column: Nonmarketed and other economic values
Carbon sequestration
Watershed protection
Non-timber forest products
Recreation and hunting
Source: Millennium Ecosystem Assessment
Total economic value
dollars per hectare
200
180
160
140
120
100
80
60
40
20
0
– 20
Portugal
Italy
Croatia
Turkey
Syria
Morocco
Tunisia
Algeria

In each case, the net benefits from the more sustainably managed ecosystem are greater than those from the converted ecosystem, even though the private (market) benefits would be greater from the converted ecosystem. (Where ranges of values are given in the original source, lower estimates are plotted here.)

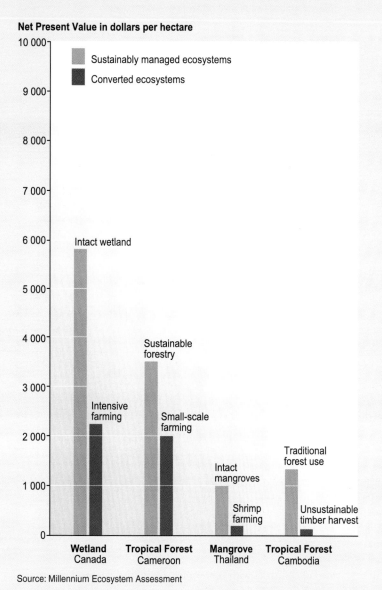

Net Present Value in dollars per hectare

Source: Millennium Ecosystem Assessment

aesthetically pleasing landscape, there is no market for these services and no one person has an incentive to pay to maintain the good. And when an action results in the degradation of a service that harms other individuals, no market mechanism exists (nor, in many cases, could it exist) to ensure that the individuals harmed are compensated for the damages they suffer.

Wealthy populations cannot be insulated from the degradation of ecosystem services. [3] Agriculture, fisheries, and forestry once formed the bulk of national economies, and the control of natural resources dominated policy agendas. But while these natural resource industries are often still important, the relative economic and political significance of other industries in industrial countries has grown over the past century as a result of the ongoing transition from agricultural to industrial and service economies, urbanization, and the development of new technologies to increase the production of some services and provide substitutes for others. Nevertheless, the degradation of ecosystem services influences human well-being in industrial regions and among wealthy populations in developing countries in many ways:

■ The physical, economic, or social impacts of ecosystem service degradation may cross boundaries. (See Figure 10.) For example, land degradation and associated dust storms or fires in one country can degrade air quality in other countries nearby.

■ Degradation of ecosystem services exacerbates poverty in developing countries, which can affect neighboring industrial countries by slowing regional economic growth and contributing to the outbreak of conflicts or the migration of refugees.

■ Changes in ecosystems that contribute to greenhouse gas emissions contribute to global climate changes that affect all countries.

■ Many industries still depend directly on ecosystem services. The collapse of fisheries, for example, has harmed many communities in industrial countries. Prospects for the forest, agriculture, fishing, and ecotourism industries are all directly tied to ecosystem services, while other sectors such as insurance, banking, and health are strongly, if less directly, influenced by changes in ecosystem services.

■ Wealthy populations of people are insulated from the harmful effects of some aspects of ecosystem degradation, but not all. For example, substitutes are typically not available when cultural services are lost.

■ Even though the relative economic importance of agriculture, fisheries, and forestry is declining in industrial countries, the importance of other ecosystem services such as aesthetic enjoyment and recreational options is growing.

While degradation of some services may sometimes be warranted to produce a greater gain in other services, often more degradation of ecosystem services takes place than is in society's interests because many of the services degraded are "public goods." [3] Although people benefit from ecosystem services such as the regulation of air and water quality or the presence of an

It is difficult to assess the implications of ecosystem changes and to manage ecosystems effectively because many of the effects are slow to become apparent, because they may be expressed primarily at some distance from where the ecosystem was changed, and because the costs and benefits of changes often accrue to different sets of stakeholders. [7] Substantial inertia (delay in the response of a system to a disturbance) exists in ecological systems. As a result, long time lags often occur between a change in a driver and the time when the full consequences of that change become apparent. For example, phosphorus is accumulating in large quantities in many agricultural soils, threatening rivers, lakes, and coastal oceans with increased eutrophication. But it may take years or decades for the full impact of the phosphorus to become apparent through erosion and other processes. Similarly, it will take centuries for global temperatures to reach equilibrium with changed concentrations of greenhouse gases in the atmosphere and even more time for biological systems to respond to the changes in climate.

Moreover, some of the impacts of ecosystem changes may be experienced only at some distance from where the change occurred. For example, changes in upstream catchments affect water flow and water quality in downstream regions; similarly, the loss of an important fish nursery area in a coastal wetland may diminish fish catch some distance away. Both the inertia in ecological systems and the temporal and spatial separation of costs and benefits of ecosystem changes often result in situations where the individuals experiencing harm from ecosystem changes (future generations, say, or downstream landowners) are not the same as the individuals gaining the benefits. These temporal and spatial patterns make it extremely difficult to fully assess costs and benefits associated with ecosystem changes or to attribute costs and benefits to different stakeholders. Moreover, the institutional arrangements now in place to manage ecosystems are poorly designed to cope with these challenges.

Increased Likelihood of Nonlinear (Stepped) and Potentially Abrupt Changes in Ecosystems

There is *established but incomplete* evidence that changes being made in ecosystems are increasing the likelihood of nonlinear changes in ecosystems (including accelerating, abrupt, and potentially irreversible changes), with important consequences for human well-being. [7] Changes in ecosystems generally take place gradually. Some changes are nonlinear, however: once a threshold is crossed, the system changes to a very different state. And these nonlinear changes are sometimes abrupt; they can also be large in magnitude and difficult, expensive, or impossible to reverse. Capabilities for predicting some nonlinear changes are improving, but for most ecosystems and for most potential nonlinear changes, while science can often warn of increased risks of change it cannot predict the thresholds at which the change will be encountered. Examples of large-magnitude nonlinear changes include:

■ *Disease emergence.* If, on average, each infected person infects at least one other person, then an epidemic spreads, while if the infection is transferred on average to less than one person, the epidemic dies out. During the 1997–98 El Niño, excessive flooding caused cholera epidemics in Djibouti, Somalia, Kenya, Tanzania, and Mozambique. Warming of the African Great Lakes due to climate change may create conditions that increase the risk of cholera transmission in the surrounding countries.

■ *Eutrophication and hypoxia.* Once a threshold of nutrient loading is achieved, changes in freshwater and coastal ecosystems can be abrupt and extensive, creating harmful algal blooms (including blooms of toxic species) and sometimes leading to the formation of oxygen-depleted zones, killing most animal life.

Figure 10. Dust Cloud off the Northwest Coast of Africa, March 6, 2004

In this image, the storm covers about one fifth of Earth's circumference. The dust clouds travel thousands of kilometers and fertilize the water off the west coast of Florida with iron. This has been linked to blooms of toxic algae in the region and respiratory problems in North America and has affected coral reefs in the Caribbean. Degradation of drylands exacerbates problems associated with dust storms.

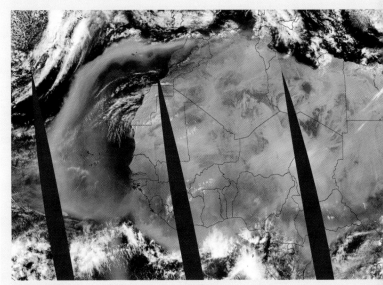

Source: National Aeronautics and Space Administration, Earth Observatory

■ *Fisheries collapse.* For example, the Atlantic cod stocks off the east coast of Newfoundland collapsed in 1992, forcing the closure of the fishery after hundreds of years of exploitation. (See Figure 11.) Most important, depleted stocks may take years to recover, or not recover at all, even if harvesting is significantly reduced or eliminated entirely.

■ *Species introductions and losses.* The introduction of the zebra mussel into aquatic systems in the United States, for instance, resulted in the extirpation of native clams in Lake St. Clair and annual costs of $100 million to the power industry and other users.

■ *Regional climate change.* Deforestation generally leads to decreased rainfall. Since forest existence crucially depends on rainfall, the relationship between forest loss and precipitation decrease can form a positive feedback, which, under certain conditions, can lead to a nonlinear change in forest cover.

The growing bushmeat trade poses particularly significant threats associated with nonlinear changes, in this case accelerating rates of change. [7] Growth in the use and trade of bushmeat is placing increasing pressure on many species, especially in Africa and Asia. While the population size of harvested species may decline gradually with increasing harvest for some time, once the harvest exceeds sustainable levels, the rate of decline of populations of the harvested species will tend to accelerate. This could place them at risk of extinction and also reduce the food supply of people dependent on these resources in the longer term. At the same time, the bushmeat trade involves relatively high levels of interaction between humans and some relatively closely related wild animals that are eaten. Again, this increases the risk of a nonlinear change, in this case the emergence of new and serious pathogens. Given the speed and magnitude of international travel today, new pathogens could spread rapidly around the world.

The increased likelihood of these nonlinear changes stems from the loss of biodiversity and growing pressures from multiple direct drivers of ecosystem change. [7] The loss of species and genetic diversity decreases the resilience of ecosystems, which is the level of disturbance that an ecosystem can undergo without crossing a threshold to a different structure or functioning. In addition, growing pressures from drivers such as overharvesting, climate change, invasive species, and nutrient loading push ecosystems toward thresholds that they might otherwise not encounter.

Exacerbation of Poverty for Some Individuals and Groups of People and Contribution to Growing Inequities and Disparities across Groups of People

Despite the progress achieved in increasing the production and use of some ecosystem services, levels of poverty remain high, inequities are growing, and many people still do not have a sufficient supply of or access to ecosystem services. [3]

■ In 2001, 1.1 billion people survived on less than $1 per day of income, with roughly 70% of them in rural areas where they are highly dependent on agriculture, grazing, and hunting for subsistence.

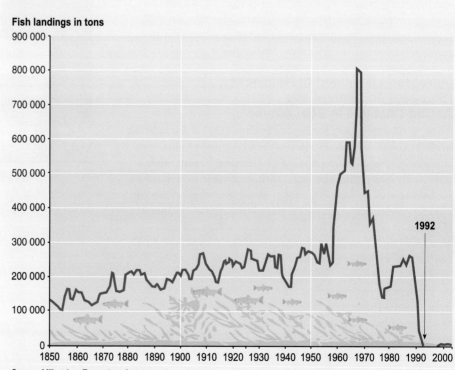

Figure 11. COLLAPSE OF ATLANTIC COD STOCKS OFF THE EAST COAST OF NEWFOUNDLAND IN 1992 (CF Box 2.4)

This collapse forced the closure of the fishery after hundreds of years of exploitation. Until the late 1950s, the fishery was exploited by migratory seasonal fleets and resident inshore small-scale fishers. From the late 1950s, offshore bottom trawlers began exploiting the deeper part of the stock, leading to a large catch increase and a strong decline in the underlying biomass. Internationally agreed quotas in the early 1970s and, following the declaration by Canada of an Exclusive Fishing Zone in 1977, national quota systems ultimately failed to arrest and reverse the decline. The stock collapsed to extremely low levels in the late 1980s and early 1990s, and a moratorium on commercial fishing was declared in June 1992. A small commercial inshore fishery was reintroduced in 1998, but catch rates declined and the fishery was closed indefinitely in 2003.

Source: Millennium Ecosystem Assessment

■ Inequality in income and other measures of human well-being has increased over the past decade. A child born in sub-Saharan Africa is 20 times more likely to die before age 5 than a child born in an industrial country, and this disparity is higher than it was a decade ago. During the 1990s, 21 countries experienced declines in their rankings in the Human Development Index (an aggregate measure of economic well-being, health, and education); 14 of them were in sub-Saharan Africa.

■ Despite the growth in per capita food production in the past four decades, an estimated 852 million people were undernourished in 2000–02, up 37 million from the period 1997–99. South Asia and sub-Saharan Africa, the regions with the largest numbers of undernourished people, are also the regions where growth in per capita food production has lagged the most. Most notably, per capita food production has declined in sub-Saharan Africa.

■ Some 1.1 billion people still lack access to improved water supply, and more than 2.6 billion lack access to improved sanitation. Water scarcity affects roughly 1–2 billion people worldwide. Since 1960, the ratio of water use to accessible supply has grown by 20% per decade.

The degradation of ecosystem services is harming many of the world's poorest people and is sometimes the principal factor causing poverty. [3, 6]

■ Half the urban population in Africa, Asia, Latin America, and the Caribbean suffers from one or more diseases associated with inadequate water and sanitation. Worldwide, approximately 1.7 million people die annually as a result of inadequate water, sanitation, and hygiene.

■ The declining state of capture fisheries is reducing an inexpensive source of protein in developing countries. Per capita fish consumption in developing countries, excluding China, declined between 1985 and 1997.

■ Desertification affects the livelihoods of millions of people, including a large portion of the poor in drylands.

The pattern of "winners" and "losers" associated with ecosystem changes—and in particular the impact of ecosystem changes on poor people, women, and indigenous peoples— has not been adequately taken into account in management decisions. [3, 6] Changes in ecosystems typically yield benefits for some people and exact costs on others who may either lose access to resources or livelihoods or be affected by externalities associated with the change. For several reasons, groups such as the poor, women, and indigenous communities have tended to be harmed by these changes.

■ Many changes in ecosystem management have involved the privatization of what were formerly common pool resources. Individuals who depended on those resources (such as indigenous peoples, forest-dependent communities, and other groups relatively marginalized from political and economic sources of power) have often lost rights to the resources.

■ Some of the people and places affected by changes in ecosystems and ecosystem services are highly vulnerable and poorly equipped to cope with the major changes in ecosystems that may occur. Highly vulnerable groups include those whose needs for ecosystem services already exceed the supply, such as people lacking adequate clean water supplies, and people living in areas with declining per capita agricultural production.

■ Significant differences between the roles and rights of men and women in many societies lead to increased vulnerability of women to changes in ecosystem services.

■ The reliance of the rural poor on ecosystem services is rarely measured and thus typically overlooked in national statistics and poverty assessments, resulting in inappropriate strategies that do not take into account the role of the environment in poverty reduction. For example, a recent study that synthesized data from 17 countries found that 22% of household income for rural communities in forested regions comes from sources typically not included in national statistics, such as harvesting wild food, fuelwood, fodder, medicinal plants, and timber. These activities generated a much higher proportion of poorer families' total income than of wealthy families', and this income was of particular significance in periods of both predictable and unpredictable shortfalls in other livelihood sources.

Development prospects in dryland regions of developing countries are especially dependent on actions to avoid the degradation of ecosystems and slow or reverse degradation where it is occurring. [3, 5] Dryland systems cover about 41% of Earth's land surface and more than 2 billion people inhabit them, more than 90% of whom are in developing countries. Dryland ecosystems (encompassing both rural and urban regions of drylands) experienced the highest population growth rate in the 1990s of any of the systems examined in the MA. (See Figure 12.) Although drylands are home to about one third of the human population, they have only 8% of the world's renewable water supply. Given the low and variable rainfall, high temperatures, low soil organic matter, high costs of delivering services such as electricity or piped water, and limited investment in infrastructure due to the low population density, people living in drylands face many challenges. They also tend to have the lowest levels of human well-being, including the lowest per capita GDP and the highest infant mortality rates.

The combination of high variability in environmental conditions and relatively high levels of poverty leads to situations where people can be highly vulnerable to changes in ecosystems, although the presence of these conditions has led to the development of very resilient land management strategies. Pressures on dryland ecosystems already exceed sustainable levels for some ecosystem services, such as soil formation and water supply, and are growing. Per capita water availability is currently only two thirds of the level required for minimum levels of human well-being. Approximately 10–20% of the world's drylands are degraded (*medium certainty*) directly harming the people living in these areas and indirectly harming a larger population through biophysical impacts (dust storms, greenhouse gas emissions, and regional climate change) and through socioeconomic impacts

MA systems with the lowest net primary productivity and lowest GDP tended to have the highest population growth rates between 1990 and 2000. Urban, inland water, and marine systems are not included due to the somewhat arbitrary nature of determining net primary productivity of the system (urban) or population growth and GDP (freshwater and marine) for them.

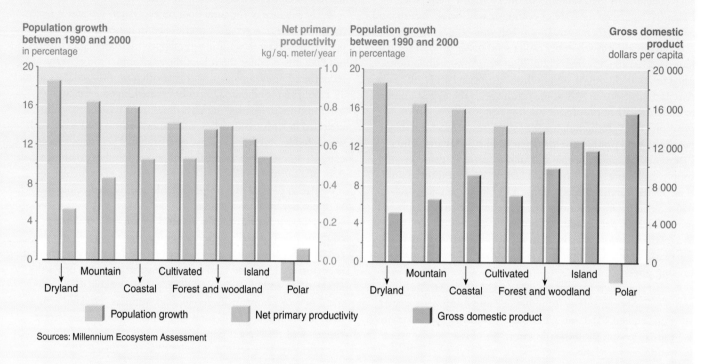

Sources: Millennium Ecosystem Assessment

(human migration and deepening poverty sometimes contributing to conflict and instability). Despite these tremendous challenges, people living in drylands and their land management systems have a proven resilience and the capability of preventing land degradation, although this can be either undermined or enhanced by public policies and development strategies.

Finding #3: *The degradation of ecosystem services could grow significantly worse during the first half of this century and is a barrier to achieving the Millennium Development Goals.*

The MA developed four scenarios to explore plausible futures for ecosystems and human well-being. (See Box 1.) The scenarios explored two global development paths, one in which the world becomes increasingly globalized and the other in which it becomes increasingly regionalized, as well as two different approaches to ecosystem management, one in which actions are reactive and most problems are addressed only after they become obvious and the other in which ecosystem management is proactive and policies deliberately seek to maintain ecosystem services for the long term.

Most of the direct drivers of change in ecosystems currently remain constant or are growing in intensity in most ecosystems. (See Figure 13.) In all four MA scenarios, the pressures on ecosystems are projected to continue to grow during the first half of this century. [4, 5] The most important direct drivers of change in ecosystems are habitat change (land use change and physical modification of rivers or water withdrawal from rivers), overexploitation, invasive alien species, pollution, and climate change. These direct drivers are often synergistic. For example, in some locations land use change can result in greater nutrient loading (if the land is converted to high-intensity agriculture), increased emissions of greenhouse gases (if forest is cleared), and increased numbers of invasive species (due to the disturbed habitat).

■ *Habitat transformation, particularly from conversion to agriculture:* Under the MA scenarios, a further 10–20% of grassland and forestland is projected to be converted between 2000 and 2050 (primarily to agriculture), as Figure 2 illustrated. The projected land conversion is concentrated in low-income countries and dryland regions. Forest cover is projected to continue to increase within industrial countries.

■ *Overexploitation, especially overfishing:* Over much of the world, the biomass of fish targeted in fisheries (including that of both the target species and those caught incidently) has been reduced by 90% relative to levels prior to the onset of industrial fishing, and the fish being harvested are increasingly coming from the less valuable lower trophic levels as populations of higher trophic level species are depleted, as shown in Figure 6. These pressures continue to grow in all the MA scenarios.

■ *Invasive alien species:* The spread of invasive alien species and disease organisms continues to increase because of both deliberate translocations and accidental introductions related to growing trade and travel, with significant harmful consequences to native species and many ecosystem services.

■ *Pollution, particularly nutrient loading:* Humans have already doubled the flow of reactive nitrogen on the continents, and some projections suggest that this may increase by roughly a further two thirds by 2050. (See Figure 14.) Three out of four MA scenarios project that the global flux of nitrogen to coastal ecosystems will increase by a further 10–20% by 2030 (*medium certainty*), with almost all of this increase occurring in developing countries. Excessive flows of nitrogen contribute to eutrophication of freshwater and coastal marine ecosystems and acidification of freshwater and terrestrial ecosystems (with implications for biodiversity in these ecosystems). To some degree, nitrogen also plays a role in creation of ground-level ozone (which leads to loss of agricultural and forest productivity), destruction of ozone in the stratosphere (which leads to depletion of the ozone layer and increased UV-B radiation on Earth, causing increased incidence of skin cancer), and climate change. The resulting health effects include the consequences of ozone pollution on asthma and respiratory function, increased allergies and asthma due to increased pollen production, the risk of blue-baby syndrome,

---

Box 1. MA Scenarios

The MA developed four scenarios to explore plausible futures for ecosystems and human well-being based on different assumptions about driving forces of change and their possible interactions:

Global Orchestration – This scenario depicts a globally connected society that focuses on global trade and economic liberalization and takes a reactive approach to ecosystem problems but that also takes strong steps to reduce poverty and inequality and to invest in public goods such as infrastructure and education. Economic growth in this scenario is the highest of the four scenarios, while it is assumed to have the lowest population in 2050.

Order from Strength – This scenario represents a regionalized and fragmented world, concerned with security and protection, emphasizing primarily regional markets, paying little attention to public goods, and taking a reactive approach to ecosystem problems. Economic growth rates are the lowest of the scenarios (particularly low in developing countries) and decrease with time, while population growth is the highest.

Adapting Mosaic – In this scenario, regional watershed-scale ecosystems are the focus of political and economic activity. Local institutions are strengthened and local ecosystem management strategies are common; societies develop a strongly proactive approach to the management of ecosystems. Economic growth rates are somewhat low initially but

increase with time, and population in 2050 is nearly as high as in *Order from Strength*.

TechnoGarden – This scenario depicts a globally connected world relying strongly on environmentally sound technology, using highly managed, often engineered, ecosystems to deliver ecosystem services, and taking a proactive approach to the management of ecosystems in an effort to avoid problems. Economic growth is relatively high and accelerates, while population in 2050 is in the mid-range of the scenarios.

The scenarios are not predictions; instead they were developed to explore the unpredictable features of change in drivers and ecosystem services. No scenario represents business as usual, although all begin from current conditions and trends.

Both quantitative models and qualitative analyses were used to develop the scenarios. For some drivers (such as land use change and carbon emissions) and ecosystem services (water withdrawals, food production), quantitative projections were calculated using established, peer-reviewed global models. Other drivers (such as rates of technological change and economic growth), ecosystem services (particularly supporting and cultural services, such as soil formation and recreational opportunities), and human well-being indicators (such as human health and social relations) were estimated qualitatively. In general, the quantitative models used for these scenarios addressed incremen-

tal changes but failed to address thresholds, risk of extreme events, or impacts of large, extremely costly, or irreversible changes in ecosystem services. These phenomena were addressed qualitatively by considering the risks and impacts of large but unpredictable ecosystem changes in each scenario.

Three of the scenarios – *Global Orchestration*, *Adapting Mosaic*, and *TechnoGarden* incorporate significant changes in policies aimed at addressing sustainable development challenges. In *Global Orchestration* trade barriers are eliminated, distorting subsidies are removed, and a major emphasis is placed on eliminating poverty and hunger. In *Adapting Mosaic*, by 2010, most countries are spending close to 13% of their GDP on education (as compared to an average of 3.5% in 2000), and institutional arrangements to promote transfer of skills and knowledge among regional groups proliferate. In *TechnoGarden* policies are put in place to provide payment to individuals and companies that provide or maintain the provision of ecosystem services. For example, in this scenario, by 2015, roughly 50% of European agriculture, and 10% of North American agriculture is aimed at balancing the production of food with the production of other ecosystem services. Under this scenario, significant advances occur in the development of environmental technologies to increase production of services, create substitutes, and reduce harmful trade-offs.

The cell color indicates impact of each driver on biodiversity in each type of ecosystem over the past 50–100 years. High impact means that over the last century the particular driver has significantly altered biodiversity in that biome; low impact indicates that it has had little influence on biodiversity in the biome. The arrows indicate the trend in the driver. Horizontal arrows indicate a continuation of the current level of impact; diagonal and vertical arrows indicate progressively increasing trends in impact. Thus, for example, if an ecosystem had experienced a very high impact of a particular driver in the past century (such as the impact of invasive species on islands), a horizontal arrow indicates that this very high impact is likely to continue. This Figure is based on expert opinion consistent with and based on the analysis of drivers of change in the various chapters of the assessment report of the MA Condition and Trends Working Group. The Figure presents global impacts and trends that may be different from those in specific regions.

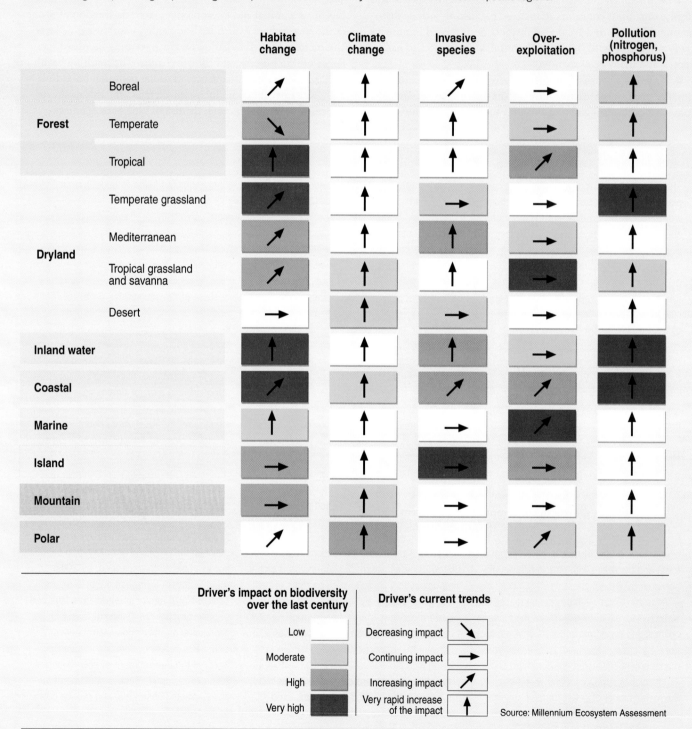

increased risk of cancer and other chronic diseases from nitrates in drinking water, and increased risk of a variety of pulmonary and cardiac diseases from the production of fine particles in the atmosphere.

■ *Anthropogenic Climate Change:* Observed recent changes in climate, especially warmer regional temperatures, have already had significant impacts on biodiversity and ecosystems, including causing changes in species distributions, population sizes, the timing of reproduction or migration events, and an increase in the frequency of pest and disease outbreaks. Many coral reefs have undergone major, although often partially reversible, bleaching episodes when local sea surface temperatures have increased during one month by 0.5–1° Celsius above the average of the hottest months

By the end of the century, climate change and its impacts may be the dominant direct driver of biodiversity loss and changes in ecosystem services globally. The scenarios developed by the Intergovernmental Panel on Climate Change project an increase in global mean surface temperature of 2.0–6.4° Celsius above preindustrial levels by 2100, increased incidence of floods and droughts, and a rise in sea level of an additional 8–88 centimeters between 1990 and 2100. Harm to biodiversity will grow worldwide with increasing rates of change in climate and increasing absolute amounts of change. In contrast, some ecosystem services in some regions may initially be enhanced by projected changes in climate (such as increases in temperature or precipitation), and thus these regions may experience net benefits at low levels of climate change. As climate change becomes more severe, however, the harmful impacts on ecosystem services outweigh the benefits in most regions of the world. The balance of scientific evidence suggests that there will be a significant net harmful impact on ecosystem services worldwide if global mean surface temperature increases more than 2° Celsius above preindustrial levels or at rates greater than 0.2° Celsius per decade (*medium certainty*). There is a wide band of uncertainty in the amount of warming that would result from any stabilized greenhouse gas concentration, but based on IPCC projections this would require an eventual CO_2 stabilization level of less than 450 parts per million carbon dioxide (*medium certainty*).

Under all four MA scenarios, the projected changes in drivers result in significant growth in consumption of ecosystem services, continued loss of biodiversity, and further degradation of some ecosystem services. [5]

■ During the next 50 years, demand for food crops is projected to grow by 70–85% under the MA scenarios, and demand for water by between 30% and 85%. Water withdrawals in developing countries are projected to increase significantly under the scenarios, although these are projected to decline in industrial countries (*medium certainty*).

■ Food security is not achieved under the MA scenarios by 2050, and child malnutrition is not eradicated (and is projected to increase in some regions in some MA scenarios) despite increasing food supply and more diversified diets (*medium certainty*).

■ A deterioration of the services provided by freshwater resources (such as aquatic habitat, fish production, and water supply for households, industry, and agriculture) is found in the scenarios, particularly in those that are reactive to environmental problems (*medium certainty*).

■ Habitat loss and other ecosystem changes are projected to lead to a decline in local diversity of native species in all four MA scenarios by 2050 (*high certainty*). Globally, the equilibrium number of plant species is projected to be reduced by roughly 10–15% as the result of habitat loss alone over the period of 1970 to 2050 in the MA scenarios (*low certainty*), and other

Figure 14. GLOBAL TRENDS IN THE CREATION OF REACTIVE NITROGEN ON EARTH BY HUMAN ACTIVITY, WITH PROJECTION TO 2050 (R9 Fig 9.1)

Most of the reactive nitrogen produced by humans comes from manufacturing nitrogen for synthetic fertilizer and industrial use. Reactive nitrogen is also created as a by-product of fossil fuel combustion and by some (nitrogen-fixing) crops and trees in agroecosystems. The range of the natural rate of bacterial nitrogen fixation in natural terrestrial ecosystems (excluding fixation in agroecosystems) is shown for comparison. Human activity now produces approximately as much reactive nitrogen as natural processes do on the continents. (Note: The 2050 projection is included in the original study and is not based on MA Scenarios.)

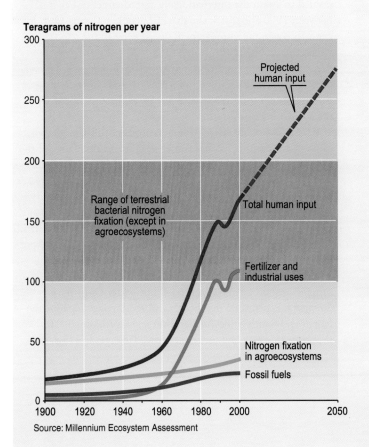

Source: Millennium Ecosystem Assessment

factors such as overharvesting, invasive species, pollution, and climate change will further increase the rate of extinction.

The degradation of ecosystem services poses a significant barrier to the achievement of the Millennium Development Goals and the MDG targets for 2015. [3] The eight Millennium Development Goals adopted by the United Nations in 2000 aim to improve human well-being by reducing poverty, hunger, child and maternal mortality, by ensuring education for all, by controlling and managing diseases, by tackling gender disparity, by ensuring environmental sustainability, and by pursuing global partnerships. Under each of the MDGs, countries have agreed to targets to be achieved by 2015. Many of the regions facing the greatest challenges in achieving these targets coincide with regions facing the greatest problems of ecosystem degradation.

Although socioeconomic policy changes will play a primary role in achieving most of the MDGs, many of the targets (and goals) are unlikely to be achieved without significant improvement in management of ecosystems. The role of ecosystem changes in exacerbating poverty (Goal 1, Target 1) for some groups of people has been described already, and the goal of environmental sustainability, including access to safe drinking water (Goal 7, Targets 9, 10, and 11), cannot be achieved as long as most ecosystem services are being degraded. Progress toward three other MDGs is particularly dependent on sound ecosystem management:

■ Hunger (Goal 1, Target 2): All four MA scenarios project progress in the elimination of hunger but at rates far slower than needed to attain the internationally agreed target of halving, between 1990 and 2015, the share of people suffering from hunger. Moreover, the improvements are slowest in the regions in which the problems are greatest: South Asia and sub-Saharan Africa. Ecosystem condition, in particular climate, soil degradation, and water availability, influences progress toward this goal through its effect on crop yields as well as through impacts on the availability of wild sources of food.

■ Child mortality (Goal 4): Undernutrition is the underlying cause of a substantial proportion of all child deaths. Three of the MA scenarios project reductions in child undernourishment by 2050 of between 10% and 60% but undernourishment increases by 10% in *Order from Strength* (*low certainty*). Child mortality is also strongly influenced by diseases associated with water quality. Diarrhea is one of the predominant causes of infant deaths worldwide. In sub-Saharan Africa, malaria additionally plays an important part in child mortality in many countries of the region.

■ Disease (Goal 6): In the more promising MA scenarios, progress toward Goal 6 is achieved, but under *Order from Strength* it is plausible that health and social conditions for the North and South could further diverge, exacerbating health problems in many low-income regions. Changes in ecosystems

influence the abundance of human pathogens such as malaria and cholera as well as the risk of emergence of new diseases. Malaria is responsible for 11% of the disease burden in Africa, and it is estimated that Africa's GDP could have been $100 billion larger in 2000 (roughly a 25% increase) if malaria had been eliminated 35 years ago. The prevalence of the following infectious diseases is particularly strongly influenced by ecosystem change: malaria, schistosomiasis, lymphatic filariasis, Japanese encephalitis, dengue fever, leishmaniasis, Chagas disease, meningitis, cholera, West Nile virus, and Lyme disease.

> **Finding #4:** *The challenge of reversing the degradation of ecosystems while meeting increasing demands for their services can be partially met under some scenarios that the MA considered, but these involve significant changes in policies, institutions, and practices that are not currently under way. Many options exist to conserve or enhance specific ecosystem services in ways that reduce negative trade-offs or that provide positive synergies with other ecosystem services.*

Three of the four MA scenarios show that significant changes in policies, institutions, and practices can mitigate many of the negative consequences of growing pressures on ecosystems, although the changes required are large and not currently under way. [5] All provisioning, regulating, and cultural ecosystem services are projected to be in worse condition in 2050 than they are today in only one of the four MA scenarios (*Order from Strength*). At least one of the three categories of services is in better condition in 2050 than in 2000 in the other three scenarios. (See Figure 15.) The scale of interventions that result in these positive outcomes are substantial and include significant investments in environmentally sound technology, active adaptive management, proactive action to address environmental problems before their full consequences are experienced, major investments in public goods (such as education and health), strong action to reduce socioeconomic disparities and eliminate poverty, and expanded capacity of people to manage ecosystems adaptively. However, even in scenarios where one or more categories of ecosystem services improve, biodiversity continues to be lost and thus the long-term sustainability of actions to mitigate degradation of ecosystem services is uncertain.

Past actions to slow or reverse the degradation of ecosystems have yielded significant benefits, but these improvements have generally not kept pace with growing pressures and demands. [8] Although most ecosystem services assessed in the MA are being degraded, the extent of that degradation would have been much greater without responses implemented in past decades. For example, more than 100,000 protected areas (including strictly protected areas such as national parks as well as areas managed for the sustainable use of natural ecosystems, including timber or wildlife harvest) covering about

The Figure shows the net change in the number of ecosystem services enhanced or degraded in the MA scenarios in each category of services for industrial and developing countries expressed as a percentage of the total number of services evaluated in that category. Thus, 100% degradation means that all the services in the category were degraded in 2050 compared with 2000, while 50% improvement could mean that three out of six services were enhanced and the rest were unchanged or that four out of six were enhanced and one was degraded. The total number of services evaluated for each category was six provisioning services, nine regulating services, and five cultural services.

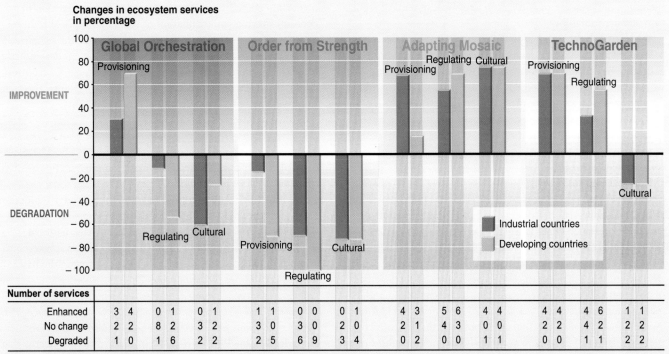

Number of services	Global Orchestration						Order from Strength						Adapting Mosaic						TechnoGarden					
Enhanced	3	4	0	1	0	1	1	1	0	0	0	1	4	3	5	6	4	4	4	4	4	6	1	1
No change	2	2	8	2	3	2	3	0	3	0	2	0	2	1	4	3	0	0	2	2	4	2	2	2
Degraded	1	0	1	6	2	2	2	5	6	9	3	4	0	2	0	0	1	1	0	0	1	1	2	2

Source: Millennium Ecosystem Assessment

11.7% of the terrestrial surface have now been established, and these play an important role in the conservation of biodiversity and ecosystem services (although important gaps in the distribution of protected areas remain, particularly in marine and freshwater systems). Technological advances have also helped lessen the increase in pressure on ecosystems caused per unit increase in demand for ecosystem services.

Substitutes can be developed for some but not all ecosystem services, but the cost of substitutes is generally high, and substitutes may also have other negative environmental consequences. [8] For example, the substitution of vinyl, plastics, and metal for wood has contributed to relatively slow growth in global timber consumption in recent years. But while the availability of substitutes can reduce pressure on specific ecosystem services, they may not always have positive net benefits on the environment. Substitution of fuelwood by fossil fuels, for example, reduces pressure on forests and lowers indoor air pollution but it also increases net greenhouse gas emissions. Substitutes are also often costlier to provide than the original ecosystem services.

Ecosystem degradation can rarely be reversed without actions that address the negative effects or enhance the positive effects of one or more of the five indirect drivers of change: population change (including growth and migration), change in economic activity (including economic growth, disparities in wealth, and trade patterns), sociopolitical factors (including factors ranging from the presence of conflict to public participation in decision-making), cultural factors, and technological change. [4] Collectively these factors influence the level of production and consumption of ecosystem services and the sustainability of the production. Both economic growth and population growth lead to increased consumption of ecosystem services, although the harmful environmental impacts of any particular level of consumption depend on the efficiency of the technologies used to produce the service. Too often, actions to slow ecosystem degradation do not address these indirect drivers. For example, forest

management is influenced more strongly by actions outside the forest sector, such as trade policies and institutions, macroeconomic policies, and policies in other sectors such as agriculture, infrastructure, energy, and mining, than by those within it.

An effective set of responses to ensure the sustainable management of ecosystems must address the indirect and drivers just described and must overcome barriers related to [8]:

■ Inappropriate institutional and governance arrangements, including the presence of corruption and weak systems of regulation and accountability.

■ Market failures and the misalignment of economic incentives.

■ Social and behavioral factors, including the lack of political and economic power of some groups (such as poor people, women, and indigenous peoples) that are particularly dependent on ecosystem services or harmed by their degradation.

■ Underinvestment in the development and diffusion of technologies that could increase the efficiency of use of ecosystem services and could reduce the harmful impacts of various drivers of ecosystem change.

■ Insufficient knowledge (as well as the poor use of existing knowledge) concerning ecosystem services and management, policy, technological, behavioral, and institutional responses that could enhance benefits from these services while conserving resources.

All these barriers are further compounded by weak human and institutional capacity related to the assessment and management of ecosystem services, underinvestment in the regulation and management of their use, lack of public awareness, and lack of awareness among decision-makers of both the threats posed by the degradation of ecosystem services and the opportunities that more sustainable management of ecosystems could provide.

The MA assessed 74 response options for ecosystem services, integrated ecosystem management, conservation and sustainable use of biodiversity, and climate change. Many of these options hold significant promise for overcoming these barriers and conserving or sustainably enhancing the supply of ecosystem services. Promising options for specific sectors are shown in Box 2, while cross-cutting responses addressing key obstacles are described in the remainder of this section.

Institutions and Governance

Changes in institutional and environmental governance frameworks are sometimes required to create the enabling conditions for effective management of ecosystems, while in other cases existing institutions could meet these needs but face significant barriers. [8] Many existing institutions at both the global and the national level have the mandate to address the degradation of ecosystem services but face a variety of challenges in doing so related in part to the need for greater cooperation across sectors and the need for coordinated responses at multiple scales.

However, since a number of the issues identified in this assessment are recent concerns and were not specifically taken into account in the design of today's institutions, changes in existing institutions and the development of new ones may sometimes be needed, particularly at the national scale.

In particular, existing national and global institutions are not well designed to deal with the management of common pool resources, a characteristic of many ecosystem services. Issues of ownership and access to resources, rights to participation in decision-making, and regulation of particular types of resource use or discharge of wastes can strongly influence the sustainability of ecosystem management and are fundamental determinants of who wins and loses from changes in ecosystems. Corruption, a major obstacle to effective management of ecosystems, also stems from weak systems of regulation and accountability.

Promising interventions include:

■ *Integration of ecosystem management goals within other sectors and within broader development planning frameworks.* The most important public policy decisions affecting ecosystems are often made by agencies and in policy arenas other than those charged with protecting ecosystems. For example, the Poverty Reduction Strategies prepared by developing-country governments for the World Bank and other institutions strongly shape national development priorities, but in general these have not taken into account the importance of ecosystems to improving the basic human capabilities of the poorest.

■ *Increased coordination among multilateral environmental agreements and between environmental agreements and other international economic and social institutions.* International agreements are indispensable for addressing ecosystem-related concerns that span national boundaries, but numerous obstacles weaken their current effectiveness. Steps are now being taken to increase the coordination among these mechanisms, and this could help to broaden the focus of the array of instruments. However, coordination is also needed between the multilateral environmental agreements and more politically powerful international institutions, such as economic and trade agreements, to ensure that they are not acting at cross-purposes. And implementation of these agreements needs to be coordinated among relevant institutions and sectors at the national level.

■ *Increased transparency and accountability of government and private-sector performance on decisions that have an impact on ecosystems, including through greater involvement of concerned stakeholders in decision-making.* Laws, policies, institutions, and markets that have been shaped through public participation in decision-making are more likely to be effective and perceived as just. Stakeholder participation also contributes to the decision-making process because it allows a better understanding of impacts and vulnerability, the distribution of costs and benefits associated with trade-offs, and the identification of a broader range of response options that are available in a specific context. And stakeholder involvement and transparency of decision-making can increase accountability and reduce corruption.

Illustrative examples of response options specific to particular sectors judged to be promising or effective are listed below. (See Appendix B.) A response is considered effective when it enhances the target ecosystem services and contributes to human well-being without significant harm to other services or harmful impacts on other groups of people. A response is considered promising if it does not have a long track record to assess but appears likely to succeed or if there are known ways of modifying the response so that it can become effective.

Agriculture
■ Removal of production subsidies that have adverse economic, social, and environmental effects.
■ Investment in, and diffusion of, agricultural science and technology that can sustain the necessary increase of food supply without harmful tradeoffs involving excessive use of water, nutrients, or pesticides.
■ Use of response polices that recognize the role of women in the production and use of food and that are designed to empower women and ensure access to and control of resources necessary for food security.
■ Application of a mix of regulatory and incentive- and market-based mechanisms to reduce overuse of nutrients.

Fisheries and Aquaculture
■ Reduction of marine fishing capacity.
■ Strict regulation of marine fisheries both regarding the establishment and implementation of quotas and steps to address unreported and unregulated harvest. Individual transferable quotas may be appropriate in some cases, particularly for cold water, single species fisheries.
■ Establishment of appropriate regulatory systems to reduce the detrimental environmental impacts of aquaculture.
■ Establishment of marine protected areas including flexible no-take zones.

Water
■ Payments for ecosystem services provided by watersheds.
■ Improved allocation of rights to freshwater resources to align incentives with conservation needs.
■ Increased transparency of information regarding water management and improved representation of marginalized stakeholders.
■ Development of water markets.
■ Increased emphasis on the use of the natural environment and measures other than dams and levees for flood control.
■ Investment in science and technology to increase the efficiency of water use in agriculture.

Forestry
■ Integration of agreed sustainable forest management practices in financial institutions, trade rules, global environment programs, and global security decision-making.
■ Empowerment of local communities in support of initiatives for sustainable use of forest products; these initiatives are collectively more significant than efforts led by governments or international processes but require their support to spread.
■ Reform of forest governance and development of country-led, strategically focused national forest programs negotiated by stakeholders.

Economics and Incentives

Economic and financial interventions provide powerful instruments to regulate the use of ecosystem goods and services. [8] Because many ecosystem services are not traded in markets, markets fail to provide appropriate signals that might otherwise contribute to the efficient allocation and sustainable use of the services. A wide range of opportunities exists to influence human behavior to address this challenge in the form of economic and financial instruments. However, market mechanisms and most economic instruments can only work effectively if supporting institutions are in place, and thus there is a need to build institutional capacity to enable more widespread use of these mechanisms.

Promising interventions include:

■ *Elimination of subsidies that promote excessive use of ecosystem services (and, where possible, transfer of these subsidies to payments for non-marketed ecosystem services).* Government subsidies paid to the agricultural sectors of OECD countries between 2001 and 2003 averaged over $324 billion annually, or one third the global value of agricultural products in 2000. A significant proportion of this total involved production subsidies that led to greater food production in industrial countries than the global market conditions warranted, promoted overuse of fertilizers and pesticides in those countries, and reduced the profitability of agriculture in developing countries. Many countries outside the OECD also have inappropriate input and production subsidies, and inappropriate subsidies are common in other sectors such as water, fisheries, and forestry. Although removal of perverse subsidies will produce net benefits, it will not be without costs. Compensatory mechanisms may be needed for poor people who are adversely affected by the removal of subsidies, and removal of agricultural subsidies within the OECD would need to be accompanied by actions designed to minimize adverse impacts on ecosystem services in developing countries.

■ *Greater use of economic instruments and market-based approaches in the management of ecosystem services.* These include:
■ Taxes or user fees for activities with "external" costs (tradeoffs not accounted for in the market). Examples include taxes on excessive application of nutrients or ecotourism user fees.

- Creation of markets, including through cap-and-trade systems. One of the most rapidly growing markets related to ecosystem services is the carbon market. Approximately 64 million tons of carbon dioxide equivalent were exchanged through projects from January to May 2004, nearly as much as during all of 2003. The value of carbon trades in 2003 was approximately $300 million. About one quarter of the trades involved investment in ecosystem services (hydropower or biomass). It is *speculated* that this market may grow to $10 billion to $44 billion by 2010. The creation of a market in the form of a nutrient trading system may also be a low-cost way to reduce excessive nutrient loading in the United States.
- Payment for ecosystem services. For example, in 1996 Costa Rica established a nationwide system of conservation payments to induce landowners to provide ecosystem services. Under this program, Costa Rica brokers contracts between international and domestic "buyers" and local "sellers" of sequestered carbon, biodiversity, watershed services, and scenic beauty. Another innovative conservation financing mechanism is "biodiversity offsets," whereby developers pay for conservation activities as compensation for unavoidable harm that a project causes to biodiversity.
- Mechanisms to enable consumer preferences to be expressed through markets. For example, current certification schemes for sustainable fisheries and forest practices provide people with the opportunity to promote sustainability through their consumer choices.

Social and Behavioral Responses

Social and behavioral responses—including population policy, public education, civil society actions, and empowerment of communities, women, and youth—can be instrumental in responding to the problem of ecosystem degradation. [8] These are generally interventions that stakeholders initiate and execute through exercising their procedural or democratic rights in efforts to improve ecosystems and human well-being.

Promising interventions include:

■ *Measures to reduce aggregate consumption of unsustainably managed ecosystem services.* The choices about what individuals consume and how much are influenced not just by considerations of price but also by behavioral factors related to culture, ethics, and values. Behavioral changes that could reduce demand for degraded ecosystem services can be encouraged through actions by governments (such as education and public awareness programs or the promotion of demand-side management), industry (commitments to use raw materials that are from sources certified as being sustainable, for example, or improved product labeling), and civil society (through raising public awareness). Efforts to reduce aggregate consumption, however, must sometimes incorporate measures to increase the access to and consumption of those same ecosystem services by specific groups such as poor people.

■ *Communication and education.* Improved communication and education are essential to achieve the objectives of environmental conventions and the Johannesburg Plan of Implementation as well as the sustainable management of natural resources more generally. Both the public and decision-makers can benefit from education concerning ecosystems and human well-being, but education more generally provides tremendous social benefits that can help address many drivers of ecosystem degradation. While the importance of communication and education is well recognized, providing the human and financial resources to undertake effective work is a continuing problem.

■ *Empowerment of groups particularly dependent on ecosystem services or affected by their degradation, including women, indigenous peoples, and young people.* Despite women's knowledge about the environment and the potential they possess, their participation in decision-making has often been restricted by economic, social, and cultural structures. Young people are also key stakeholders in that they will experience the longer-term consequences of decisions made today concerning ecosystem services. Indigenous control of traditional homelands can sometimes have environmental benefits, although the primary justification continues to be based on human and cultural rights.

Technological Responses

Given the growing demands for ecosystem services and other increased pressures on ecosystems, the development and diffusion of technologies designed to increase the efficiency of resource use or reduce the impacts of drivers such as climate change and nutrient loading are essential. [8] Technological change has been essential for meeting growing demands for some ecosystem services, and technology holds considerable promise to help meet future growth in demand. Technologies already exist for reduction of nutrient pollution at reasonable costs—including technologies to reduce point source emissions, changes in crop management practices, and precision farming techniques to help control the application of fertilizers to a field, for example—but new policies are needed for these tools to be applied on a sufficient scale to slow and ultimately reverse the increase in nutrient loading (even while increasing nutrient application in regions such as sub-Saharan Africa where too little fertilizer is being applied). However, negative impacts on ecosystems and human well-being have sometimes resulted from new technologies, and thus careful assessment is needed prior to their introduction.

Promising interventions include:

■ *Promotion of technologies that enable increased crop yields without harmful impacts related to water, nutrient, and pesticide use.* Agricultural expansion will continue to be one of the major drivers of biodiversity loss well into the twenty-first century. Development, assessment, and diffusion of technologies that could increase the production of food per unit area sustainably without harmful trade-offs related to excessive consumption of water or use of nutrients or pesticides would significantly lessen pressure on other ecosystem services.

■ *Restoration of ecosystem services.* Ecosystem restoration activities are now common in many countries. Ecosystems with some features of the ones that were present before conversion can often be established and can provide some of the original ecosystem services. However, the cost of restoration is generally extremely high compared with the cost of preventing the degradation of the ecosystem. Not all services can be restored, and heavily degraded services may require considerable time for restoration.

■ *Promotion of technologies to increase energy efficiency and reduce greenhouse gas emissions.* Significant reductions in net greenhouse gas emissions are technically feasible due to an extensive array of technologies in the energy supply, energy demand, and waste management sectors. Reducing projected emissions will require a portfolio of energy production technologies ranging from fuel switching (coal/oil to gas) and increased power plant efficiency to increased use of renewable energy technologies, complemented by more efficient use of energy in the transportation, buildings, and industry sectors. It will also involve the development and implementation of supporting institutions and policies to overcome barriers to the diffusion of these technologies into the marketplace, increased public and private-sector funding for research and development, and effective technology transfer.

Knowledge Responses

Effective management of ecosystems is constrained both by the lack of knowledge and information about different aspects of ecosystems and by the failure to use adequately the information that does exist in support of management decisions. [8, 9] In most regions, for example, relatively limited information exists about the status and economic value of most ecosystem services, and their depletion is rarely tracked in national economic accounts. Basic global data on the extent and trend in different types of ecosystems and land use are surprisingly scarce. Models used to project future environmental and economic conditions have limited capability of incorporating ecological "feedbacks," including nonlinear changes in ecosystems, as well as behavioral feedbacks such as learning that may take place through adaptive management of ecosystems.

At the same time, decision-makers do not use all of the relevant information that is available. This is due in part to institutional failures that prevent existing policy-relevant scientific information from being made available to decision-makers and in part to the failure to incorporate other forms of knowledge and information (such as traditional knowledge and practitioners' knowledge) that are often of considerable value for ecosystem management.

Promising interventions include:

■ *Incorporation of nonmarket values of ecosystems in resource management and investment decisions.* Most resource management and investment decisions are strongly influenced by considerations of the monetary costs and benefits of alternative policy choices. Decisions can be improved if they are informed by the total economic value of alternative management options and involve deliberative mechanisms that bring to bear noneconomic considerations as well.

■ *Use of all relevant forms of knowledge and information in assessments and decision-making, including traditional and practitioners' knowledge.* Effective management of ecosystems typically requires "place-based" knowledge—that is, information about the specific characteristics and history of an ecosystem. Traditional knowledge or practitioners' knowledge held by local resource managers can often be of considerable value in resource management, but it is too rarely incorporated into decision-making processes and indeed is often inappropriately dismissed.

■ *Enhancing and sustaining human and institutional capacity for assessing the consequences of ecosystem change for human well-being and acting on such assessments.* Greater technical capacity is

needed for agriculture, forest, and fisheries management. But the capacity that exists for these sectors, as limited as it is in many countries, is still vastly greater than the capacity for effective management of other ecosystem services.

A variety of frameworks and methods can be used to make better decisions in the face of uncertainties in data, prediction, context, and scale. Active adaptive management can be a particularly valuable tool for reducing uncertainty about ecosystem management decisions. [8] Commonly used decision-support methods include cost-benefit analysis, risk assessment, multicriteria analysis, the precautionary principle, and vulnerability analysis. Scenarios also provide one means to cope with many aspects of uncertainty, but our limited understanding of ecological systems and human responses shrouds any individual scenario in its own characteristic uncertainty. Active adaptive management is a tool that can be particularly valuable given the high levels of uncertainty surrounding coupled socioecological systems. This involves the design of management programs to test hypotheses about how components of an ecosystem function and interact, thereby reducing uncertainty about the system more rapidly than would otherwise occur.

Sufficient information exists concerning the drivers of change in ecosystems, the consequences of changes in ecosystem services for human well-being, and the merits of various response options to enhance decision-making in support of sustainable development at all scales. However, many research needs and information gaps were identified in this assessment, and actions to address those needs could yield substantial benefits in the form of improved information for policy and action. [9] Due to gaps in data and knowledge, this assessment was unable to answer fully a number of questions posed by its users. Some of these gaps resulted from weaknesses in monitoring systems related to ecosystem services and their linkages with human well-being. In other cases, the assessment revealed significant needs for further research, such the need to improve understanding of nonlinear changes in ecosystems and of the economic value of alternative management options. Investments in improved monitoring and research, combined with additional assessments of ecosystem services in different nations and regions, would significantly enhance the utility of any future global assessment of the consequences of ecosystem change for human well-being.

KEY QUESTIONS IN THE MILLENNIUM ECOSYSTEM ASSESSMENT

1. *How have ecosystems changed?*   **26**

2. *How have ecosystem services and their uses changed?*   **39**

3. *How have ecosystem changes affected human well-being and poverty alleviation?*   **49**

4. *What are the most critical factors causing ecosystem changes?*   **64**

5. *How might ecosystems and their services change in the future under various plausible scenarios?*   **71**

6. *What can be learned about the consequences of ecosystem change for human well-being at sub-global scales?*   **84**

7. *What is known about time scales, inertia, and the risk of nonlinear changes in ecosystems?*   **88**

8. *What options exist to manage ecosystems sustainably?*   **92**

9. *What are the most important uncertainties hindering decision-making concerning ecosystems?*   **101**

1. *How have ecosystems changed?*

Ecosystem Structure

The structure of the world's ecosystems changed more rapidly in the second half of the twentieth century than at any time in recorded human history, and virtually all of Earth's ecosystems have now been significantly transformed through human actions. The most significant change in the structure of ecosystems has been the transformation of approximately one quarter (24%) of Earth's terrestrial surface to cultivated systems (C26.1.2). (See Box 1.1.) More land was converted to cropland in the 30 years after 1950 than in the 150 years between 1700 and 1850 (C26).

Between 1960 and 2000, reservoir storage capacity quadrupled (C7.2.4); as a result, the amount of water stored behind large dams is estimated to be three to six times the amount held by natural river channels (this excludes natural lakes) (C7.3.2). (See Figure 1.1.) In countries for which sufficient multiyear data are available (encompassing more than half of the present-day mangrove area), approximately 35% of mangroves were lost in the last two decades (C19.2.1). Roughly 20% of the world's coral reefs were lost and an additional 20% degraded in the last several decades of the twentieth century (C19.2.1). Box 1.1 and Table 1.1 summarize important characteristics and trends in different ecosystems.

Although the most rapid changes in ecosystems are now taking place in developing countries, industrial countries historically experienced comparable rates of change. Croplands expanded rapidly in Europe after 1700 and in North America and the former Soviet Union particularly after 1850 (C26.1.1). Roughly 70% of the original temperate forests and grasslands and Mediterranean forests had been lost by 1950, largely through conversion to agriculture (C4.4.3). Historically, deforestation has been much more intensive in temperate regions than in the tropics, and Europe is the continent with the smallest fraction of its original forests remaining (C21.4.2). However, changes prior to the industrial era seemed to occur at much slower rates than current transformations.

The ecosystems and biomes that have been most significantly altered globally by human activity include marine and freshwater ecosystems, temperate broadleaf forests, temperate

(continued on page 32)

Figure 1.1. TIME SERIES OF INTERCEPTED CONTINENTAL RUNOFF AND LARGE RESERVOIR STORAGE, 1900–2000 (C7 Fig 7.8)

The series is taken from a subset of large reservoirs (>0.5 cubic kilometers storage each) totaling about 65% of the global total reservoir storage for which information was available that allowed the reservoir to be georeferenced to river networks and discharge. The years 1960–2000 have shown a rapid move toward flow stabilization, which has slowed recently in some parts of the world due to the growing social, economic, and environmental concerns surrounding large hydraulic engineering works.

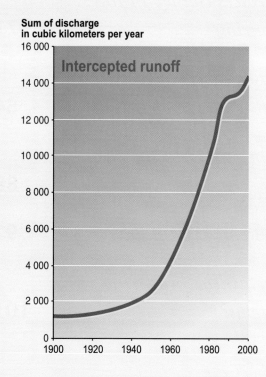

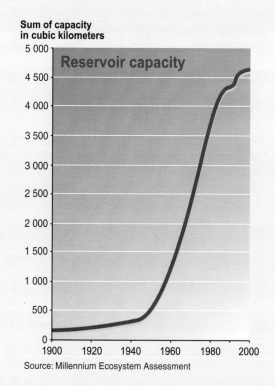

Source: Millennium Ecosystem Assessment

Box 1.1. Characteristics of the World's Ecological Systems

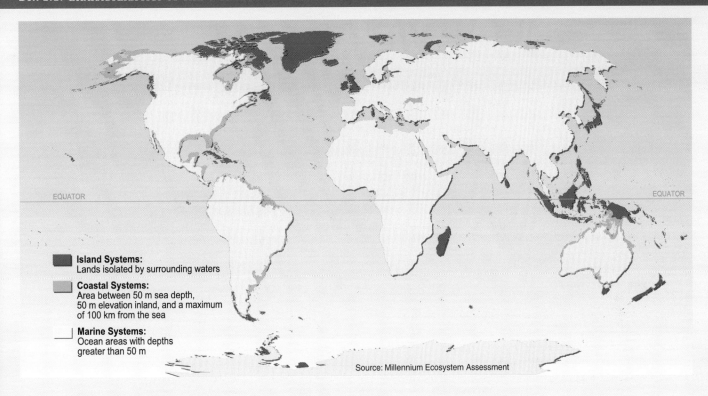

Island Systems:
Lands isolated by surrounding waters

Coastal Systems:
Area between 50 m sea depth,
50 m elevation inland, and a maximum
of 100 km from the sea

Marine Systems:
Ocean areas with depths
greater than 50 m

Source: Millennium Ecosystem Assessment

We report assessment findings for 10 categories of the land and marine surface, which we refer to as "systems": forest, cultivated, dryland, coastal, marine, urban, polar, inland water, island, and mountain. Each category contains a number of ecosystems. However, ecosystems within each category share a suite of biological, climatic, and social factors that tend to be similar within categories and differ across categories. The MA reporting categories are not spatially exclusive; their areas often overlap. For example, transition zones between forest and cultivated lands are included in both the forest system and cultivated system reporting categories. These reporting categories were selected because they correspond to the regions of responsibility of different government ministries (such as agriculture, water, forestry, and so forth) and because they are the categories used within the Convention on Biological Diversity.

Marine, Coastal, and Island Systems

■ Marine systems are the world's oceans. For mapping purposes, the map shows ocean areas where the depth is greater than 50 meters. Global fishery catches from marine systems peaked in the late 1980s and are now declining despite increasing fishing effort (C18.ES).

■ Coastal systems refer to the interface between ocean and land, extending seawards to about the middle of the continental shelf and inland to include all areas strongly influenced by proximity to the ocean. The map shows the area between 50 meters below mean sea level and 50 meters above the high tide level or extending landward to a distance 100 kilometers from shore. Coastal systems include coral reefs, intertidal zones, estuaries, coastal aquaculture, and seagrass communities. Nearly half of the world's major cities (having more than 500,000 people) are located within 50 kilometers of the coast, and coastal population densities are 2.6 times larger than the density of inland areas. By all commonly used measures, the human well-being of coastal inhabitants is on average much higher than that of inland communities (C19.3.1).

■ Islands are lands (both continental and oceanic) isolated by surrounding water and with a high proportion of coast to hinterland. For mapping purposes, the MA uses the ESRI ArcWorld Country Boundary dataset, which contains nearly 12,000 islands. Islands smaller than 1.5 hectares are not mapped or included in the statistics. The

largest island included is Greenland. The map includes islands within 2 kilometers of the mainland (e.g., Long Island in the United States), but the statistics provided for island systems in this report exclude these islands. Island states, together with their exclusive economic zones, cover 40% of the world's oceans (C23.ES). Island systems are especially sensitive to disturbances, and the majority of recorded extinctions have occurred on island systems, although this pattern is changing, and over the past 20 years as many extinctions have occurred on continents as on islands (C4.ES).

Urban, Dryland, and Polar Systems

■ Urban systems are built environments with a high human density. For mapping purposes, the MA uses known human settlements with a population of 5,000 or more, with boundaries delineated by observing persistent night-time lights or by inferring areal extent in the cases where such observations are absent. The world's urban population increased from about 200 million in 1900 to 2.9 billion in 2000, and the number of cities with populations in excess of 1 million increased from 17 in 1900 to 388 in 2000 (C27.ES).

(*continued on page 28*)

BOX 1.1. CHARACTERISTICS OF THE WORLD'S ECOLOGICAL SYSTEMS *(continued)*

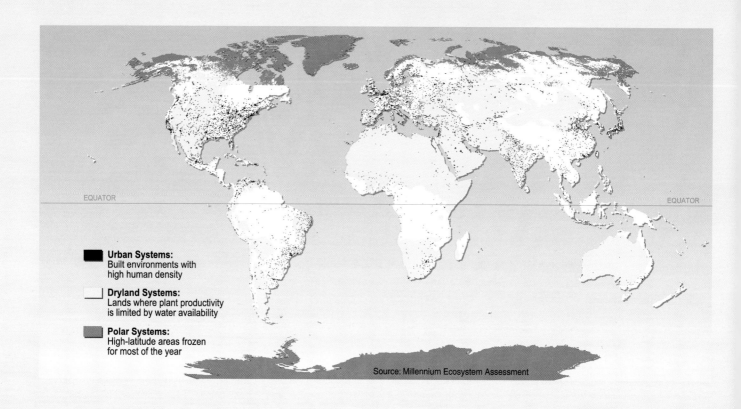

Urban Systems:
Built environments with
high human density

Dryland Systems:
Lands where plant productivity
is limited by water availability

Polar Systems:
High-latitude areas frozen
for most of the year

Source: Millennium Ecosystem Assessment

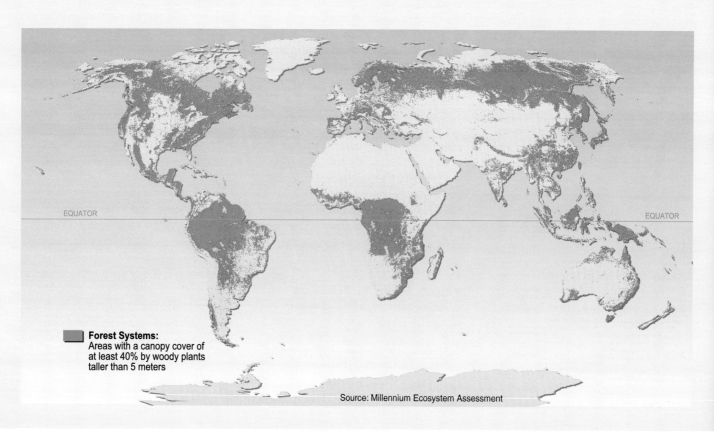

Forest Systems:
Areas with a canopy cover of
at least 40% by woody plants
taller than 5 meters

Source: Millennium Ecosystem Assessment

■ Dryland systems are lands where plant production is limited by water availability; the dominant human uses are large mammal herbivory, including livestock grazing, and cultivation. The map shows drylands as defined by the U.N. Convention to Combat Desertification, namely lands where annual precipitation is less than two thirds of potential evapotranspiration—from dry subhumid areas (ratio ranges 0.50–0.65) through semiarid, arid, and hyperarid (ratio <0.05), but excluding polar areas. Drylands include cultivated lands, scrublands, shrublands, grasslands, savannas, semi-deserts, and true deserts. Dryland systems cover about 41% of Earth's land surface and are inhabited by more than 2 billion people (about one third of the total population) (C22.ES). Croplands cover approximately 25% of drylands (C22 Table 22.2), and dryland rangelands support approximately 50% of the world's livestock (C22). The current socioeconomic condition of people in dryland systems, of which about 90% are in developing countries, is worse than in other areas. Fresh water availability in drylands is projected to be further reduced from the current average of 1,300 cubic meters per person per year in 2000, which is already below the threshold of 2,000 cubic meters required for minimum human well-being and sustainable development (C22.ES). Approximately 10–20% of the world's drylands are degraded (*medium certainty*) (C22.ES).

■ Polar systems are high-latitude systems frozen for most of the year, including ice caps, areas underlain by permafrost, tundra, polar deserts, and polar coastal areas. Polar systems do not include high-altitude cold systems in low latitudes. Temperature in polar systems is on average warmer now than at any time in the last 400 years, resulting in widespread thaw of permafrost and reduction of sea ice (C25.ES). Most changes in feedback processes that occur in polar regions magnify trace gas–induced global warming trends and reduce the capacity of polar regions to act as a cooling system for Earth (C25.ES). Tundra constitutes the largest natural wetland in the world (C25.1).

Forest Systems

■ Forest systems are lands dominated by trees; they are often used for timber, fuelwood, and non-wood forest products. The map shows areas with a canopy cover of at least 40% by woody plants taller than 5 meters. Forests include temporarily cut-over forests and plantations but exclude orchards and agroforests where the main products are food crops. The global area of forest systems has been reduced by one half over the past three centuries. Forests have effectively disappeared in 25 countries, and another 29 have lost more than 90% of their forest cover (C21.ES). Forest systems are associated with the regulation of 57% of total water run-off. About 4.6 billion people depend for all or some of their water on supplies from forest systems (C7 Table 7.2). From 1990 to 2000, the global area of temperate forest increased by almost 3 million hectares per year, while deforestation in the tropics occured at an average rate exceeding 12 million hectares per year over the past two decades (C.SDM).

Cultivated Systems

■ Cultivated systems are lands dominated by domesticated species and used for and substantially changed by crop, agroforestry, or aquaculture production. The map shows areas in which at least 30% by area of the landscape comes under cultivation in any particular year. Cultivated systems, including croplands,

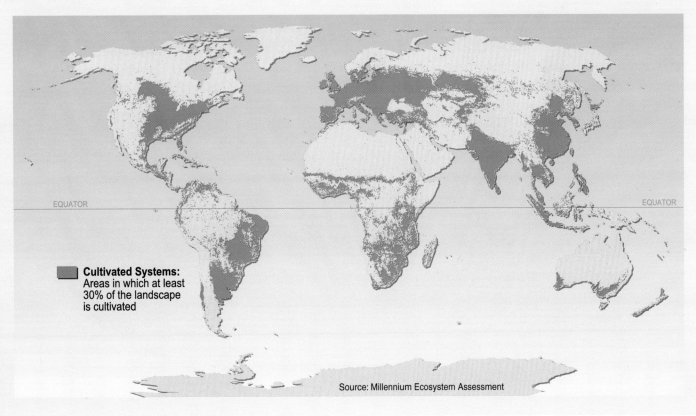

Cultivated Systems: Areas in which at least 30% of the landscape is cultivated

Source: Millennium Ecosystem Assessment

(*continued on page 30*)

Box 1.1. Characteristics of the World's Ecological Systems *(continued)*

shifting cultivation, confined livestock production, and freshwater aquaculture, cover approximately 24% of total land area. In the last two decades, the major areas of cropland expansion were located in Southeast Asia, parts of South Asia, the Great Lakes region of eastern Africa, the Amazon Basin, and the U.S. Great Plains. The major decreases of cropland occurred in the southeastern United States, eastern China, and parts of Brazil and Argentina (C26.1.1). Most of the increase in food demand of the past 50 years has been met by intensification of crop, livestock, and aquaculture systems rather than expansion of production area. In developing countries, over the period 1961–99 expansion of harvested land contributed only 29% to growth in crop production, although in sub-Saharan Africa expansion accounted for two thirds of growth in production (C26.1.1). Increased yields of crop production systems have reduced the pressure to convert natural ecosystems into cropland, but intensification has increased pressure on inland water ecosystems, generally reduced biodiversity within agricultural

landscapes, and it requires higher energy inputs in the form of mechanization and the production of chemical fertilizers. Cultivated systems provide only 16% of global runoff, although their close proximity to humans means that about 5 billion people depend for all or some of their water on supplies from cultivated systems (C7 Table 7.2). Such proximity is associated with nutrient and industrial water pollution.

Inland Water and Mountain Systems

■ Inland water systems are permanent water bodies inland from the coastal zone and areas whose properties and use are dominated by the permanent, seasonal, or intermittent occurrence of flooded conditions. Inland waters include rivers, lakes, floodplains, reservoirs, wetlands, and inland saline systems. (Note that the wetlands definition used by the Ramsar Convention includes the MA inland water and coastal system categories.) The biodiversity of inland waters appears to be in a worse condition than that of any other system, driven by declines in both the area of wetlands and the water quality in inland waters (C4 and

C20). It is *speculated* that 50% of inland water area (excluding large lakes) has been lost globally (C20.ES). Dams and other infrastructure fragment 60% of the large river systems in the world (C20.4.2).

■ Mountain systems are steep and high lands. The map is based on elevation and, at lower elevations, a combination of elevation, slope, and local topography. Some 20% (or 1.2 billion) of the world's people live in mountains or at their edges, and half of humankind depends, directly or indirectly, on mountain resources (largely water) (C24.ES). Nearly all—90%—of the 1.2 billion people in mountains live in countries with developing or transition economies. In these countries, 7% of the total mountain area is currently classified as cropland, and people are often highly dependent on local agriculture or livestock production (C24.3.2). About 4 billion people depend for all or some of their water on supplies from mountain systems. Some 90 million mountain people—almost all those living above 2,500 meters—live in poverty and are considered especially vulnerable to food insecurity (C24.1.4).

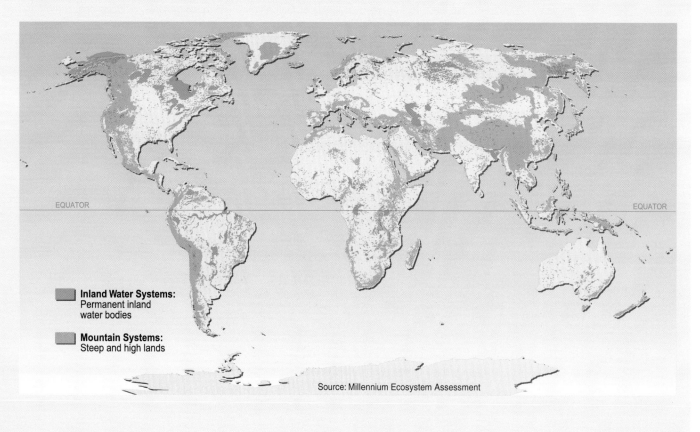

EQUATOR EQUATOR

Inland Water Systems:
Permanent inland
water bodies

Mountain Systems:
Steep and high lands

Source: Millennium Ecosystem Assessment

Table 1.1. Comparative Table of Systems as Reported by the Millennium Ecosystem Assessment (C.SDM)

Note that as described in Box 1.1, the boundaries of these systems often overlap. Statistics for different systems can therefore be compared but cannot be totaled across systems, as this would result in partial double-counting.

System and Subsystem	Area[a] (million sq. km.)	Share of Terrestrial Surface of Earth (percent)	Population Density (people per sq. km.) Urban	Population Density (people per sq. km.) Rural	Growth Rate (percent 1990–2000)	GDP per Capita (dollars)	Infant Mortality Rate[b] (deaths per 1,000 live births)	Mean NPP (kg. carbon per sq. meter per year)	Share of System Covered by PAs[c] (percent)	Share of Area Transformed[d] (percent)
Marine	**349.3**	**68.6**[e]	–	–	–	–	–	**0.15**	**0.3**	–
Coastal	**17.2**	**4.1**	**1,105**	**70**	**15.9**	**8,960**	**41.5**	–	**7**	–
Terrestrial	6.0	4.1	1,105	70	15.9	8,960	41.5	0.52	4	11
Marine	11.2	2.2[e]	–	–	–	–	–	0.14	9	–
Inland water[f]	**10.3**	**7.0**	**817**	**26**	**17.0**	**7,300**	**57.6**	**0.36**	**12**	**11**
Forest/woodland	**41.9**	**28.4**	**472**	**18**	**13.5**	**9,580**	**57.7**	**0.68**	**10**	**42**
Tropical/sub-tropical	23.3	15.8	565	14	17.0	6,854	58.3	0.95	11	34
Temperate	6.2	4.2	320	7	4.4	17,109	12.5	0.45	16	67
Boreal	12.4	8.4	114	0.1	–3.7	13,142	16.5	0.29	4	25
Dryland	**59.9**	**40.6**	**750**	**20**	**18.5**	**4,930**	**66.6**	**0.26**	**7**	**18**
Hyperarid	9.6	6.5	1,061	1	26.2	5,930	41.3	0.01	11	1
Arid	15.3	10.4	568	3	28.1	4,680	74.2	0.12	6	5
Semiarid	22.3	15.3	643	10	20.6	5,580	72.4	0.34	6	25
Dry subhumid	12.7	8.6	711	25	13.6	4,270	60.7	0.49	7	35
Island	**7.1**	**4.8**	**1,020**	**37**	**12.3**	**11,570**	**30.4**	**0.54**	**17**	**17**
Island states	4.7	3.2	918	14	12.5	11,148	30.6	0.45	18	21
Mountain	**35.8**	**24.3**	**63**	**3**	**16.3**	**6,470**	**57.9**	**0.42**	**14**	**12**
300–1,000m	13.0	8.8	58	3	12.7	7,815	48.2	0.47	11	13
1,000–2,500m	11.3	7.7	69	3	20.0	5,080	67.0	0.45	14	13
2,500–4,500m	9.6	6.5	90	2	24.2	4,144	65.0	0.28	18	6
> 4,500m	1.8	1.2	104	0	25.3	3,663	39.4	0.06	22	0.3
Polar	**23.0**	**15.6**	**161**[g]	**0.06**[g]	**–6.5**	**15,401**	**12.8**	**0.06**	**42**[g]	**0.3**[g]
Cultivated	**35.3**	**23.9**	**786**	**70**	**14.1**	**6,810**	**54.3**	**0.52**	**6**	**47**
Pasture	0.1	0.1	419	10	28.8	15,790	32.8	0.64	4	11
Cropland	8.3	5.7	1,014	118	15.6	4,430	55.3	0.49	4	62
Mixed (crop and other)	26.9	18.2	575	22	11.8	11,060	46.5	0.6	6	43
Urban	**3.6**	**2.4**	**681**	–	**12.7**	**12,057**	**36.5**	**0.47**	**0**	**100**
GLOBAL	**510**	–	**681**	**13**	**16.7**	**7,309**	**57.4**	–	**4**	**38**

[a] Area estimates based on GLC2000 dataset for the year 2000 except for cultivated systems where area is based on GLCCD v2 dataset for the years 1992–1993 (C26 Box1).

[b] Deaths of children less than one year old per 1,000 live births.

[c] Includes only natural protected areas in IUCN categories I to VI.

[d] For all systems except forest/woodland, area transformed is calculated from land depicted as cultivated or urban areas by GLC2000 land cover data set. The area transformed for forest/woodland systems is calculated as the percentage change in area between potential vegetation (forest biomes of the WWF ecoregions) and current forest/woodland areas in GLC2000. Note: 22 percent of the forest/woodland system falls outside forest biomes and is therefore not included in this analysis.

[e] Percent of total surface of Earth.

[f] Population density, growth rate, GDP per capita, and growth rate for the inland water system have been calculated with an area buffer of 10 kilometers.

[g] Excluding Antarctica.

grasslands, Mediterranean forests, and tropical dry forests. (See Figure 1.2 and C18, C20.) Within marine systems, the world's demand for food and animal feed over the last 50 years has resulted in fishing pressure so strong that the biomass of both targeted species and those caught incidentally (the "bycatch") has been reduced in much of the world to one tenth of the levels prior to the onset of industrial fishing (C18.ES). Globally, the degradation of fisheries is also reflected in the fact that the fish being harvested are increasingly coming from the less valuable lower trophic levels as populations of higher trophic level species are depleted. (See Figure 1.3.)

Freshwater ecosystems have been modified through the creation of dams and through the withdrawal of water for human use. The construction of dams and other structures along rivers has moderately or strongly affected flows in 60% of the large river systems in the world (C20.4.2). Water removal for human uses has reduced the flow of several major rivers, including the Nile, Yellow, and Colorado Rivers, to the extent that they do not always flow to the sea. As water flows have declined, so have sediment flows, which are the source of nutrients important for the maintenance of estuaries. Worldwide, although human activities have increased sediment flows in rivers by about 20%, reservoirs and water diversions prevent about 30% of sediments from reaching the oceans, resulting in a net reduction of sediment delivery to estuaries of roughly 10% (C19.ES).

Within terrestrial ecosystems, more than two thirds of the area of 2 of the world's 14 major terrestrial biomes (temperate grasslands and Mediterranean forests) and more than half of the area of 4 other biomes (tropical dry forests, temperate broadleaf forests, tropical grassland, and flooded grasslands) had been converted (primarily to agriculture) by 1990, as Figure 1.3 indicated. Among the major biomes, only tundra and boreal forests show negligible levels of loss and conversion, although they have begun to be affected by climate change.

Globally, the rate of conversion of ecosystems has begun to slow largely due to reductions in the rate of expansion of cultivated land, and in some regions (particularly in

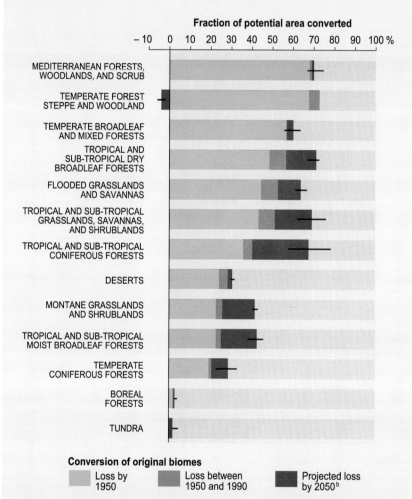

Figure 1.2. CONVERSION OF TERRESTRIAL BIOMES[a]
(Adapted from C4, S10)

It is not possible to estimate accurately the extent of different biomes prior to significant human impact, but it is possible to determine the "potential" area of biomes based on soil and climatic conditions. This Figure shows how much of that potential area is estimated to have been converted by 1950 (*medium certainty*), how much was converted between 1950 and 1990 (*medium certainty*), and how much would be converted under the four MA scenarios (*low certainty*) between 1990 and 2050. Mangroves are not included here because the area was too small to be accurately assessed. Most of the conversion of these biomes is to cultivated systems.

Conversion of original biomes

Loss by 1950 | Loss between 1950 and 1990 | Projected loss by 2050[b]

[a] A biome is the largest unit of ecological classification that is convenient to recognize below the entire globe, such as temperate broadleaf forests or montane grasslands. A biome is a widely used ecological categorization, and because considerable ecological data have been reported and modeling undertaken using this categorization, some information in this assessment can only be reported based on biomes. Whenever possible, however, the MA reports information using 10 socioecological systems, such as forest, cultivated, coastal, and marine, because these correspond to the regions of responsibility of different government ministries and because they are the categories used within the Convention on Biological Diversity.

[b] According to the four MA scenarios. For 2050 projections, the average value of the projections under the four scenarios is plotted and the error bars (black lines) represent the range of values from the different scenarios.

Source: Millennium Ecosystem Assessment

temperate zones) ecosystems are returning to conditions and species compositions similar to their pre-conversion states. Yet rates of ecosystem conversion remain high or are increasing for specific ecosystems and regions. Under the aegis of the MA, the first systematic examination of the status and trends in terrestrial and coastal land cover was carried out using global and regional datasets. The pattern of deforestation, afforestation, and dryland degradation between 1980 and 2000 is shown in Figure 1.4. Opportunities for further expansion of cultivation are diminishing in many regions of the world as most of the land well-suited for intensive agriculture has been converted to cultivation (C26. ES). Increased agricultural productivity is also diminishing the need for agricultural expansion.

As a result of these two factors, a greater fraction of land in cultivated systems (areas with at least 30% of land cultivated) is actually being cultivated, the intensity of cultivation of land is increasing, fallow lengths are decreasing, and management practices are shifting from monocultures to polycultures. Since 1950, cropland areas have stabilized in North America and decreased in Europe and China (C26.1.1). Cropland areas in the Former Soviet Union have decreased since 1960 (C26.1.1). Within temperate and boreal zones, forest cover increased by approximately 2.9 million hectares per year in the 1990s, of which approximately 40% was forest plantations (C21.4.2). In some cases, rates of conversion of ecosystems have apparently slowed because most of the ecosystem has now been converted, as is the case with temperate broadleaf forests and Mediterranean forests (C4.4.3)

Ecosystem Processes

Ecosystem processes, including water, nitrogen, carbon, and phosphorus cycling, changed more rapidly in the second half of the twentieth century than at any time in recorded human history. Human modifications of ecosystems have changed not only the structure of the systems (such as what habitats or species are present in a particular location), but their processes and functioning as well. The capacity of ecosystems to provide services derives directly from the operation of natural biogeochemical cycles that in some cases have been significantly modified.

■ *Water Cycle:* Water withdrawals from rivers and lakes for irrigation or for urban or industrial use doubled between 1960 and 2000 (C7.2.4). (Worldwide, 70% of water use is for agriculture (C7.2.2).) Large reservoir construction has doubled or tripled the residence time of river water—the average time, that is, that a drop of water takes to reach the sea (C7.3.2). Globally, humans use slightly more than 10% of the available renewable freshwater supply through household, agricultural, and industrial activities (C7.2.3), although in some regions such as the Middle East and North Africa, humans use 120% of renewable supplies (the excess is obtained through the use of groundwater supplies at rates greater than their rate of recharge) (C7.2.2).

■ *Carbon Cycle:* Since 1750, the atmospheric concentration of carbon dioxide has increased by about 34% (from about 280 parts per million to 376 parts per million in 2003) (S7.3.1). Approximately 60% of that increase (60 parts per million) has taken place since 1959. The effect of changes in terrestrial

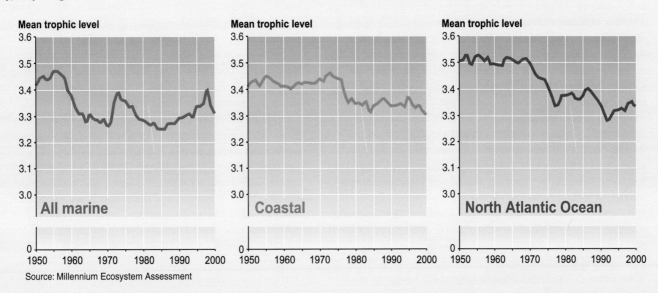

Figure 1.3. DECLINE IN TROPHIC LEVEL OF FISHERIES CATCH SINCE 1950 (C18)

A trophic level of an organism is its position in a food chain. Levels are numbered according to how far particular organisms are along the chain from the primary producers at level 1, to herbivores (level 2), to predators (level 3), to carnivores or top carnivores (level 4 or 5). Fish at higher trophic levels are typically of higher economic value. The decline in the trophic level harvested is largely a result of the overharvest of fish at higher trophic levels.

Source: Millennium Ecosystem Assessment

In the case of forest cover change, the studies refer to the period 1980–2000 and are based on national statistics, remote sensing, and to a limited degree expert opinion. In the case of land cover change resulting from degradation in drylands (desertification), the period is unspecified but inferred to be within the last half-century, and the major study was entirely based on expert opinion, with associated *low certainty*. Change in cultivated area is not shown.

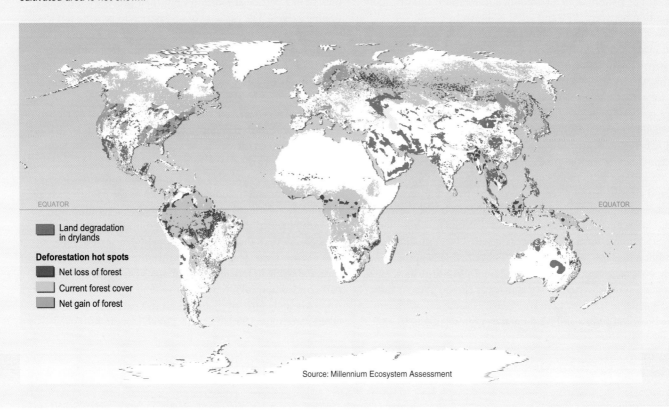

EQUATOR   EQUATOR

Land degradation
in drylands

Deforestation hot spots

Net loss of forest

Current forest cover

Net gain of forest

Source: Millennium Ecosystem Assessment

ecosystems on the carbon cycle reversed during the last 50 years. Those ecosystems were on average a net source of CO_2 during the nineteenth and early twentieth centuries (primarily due to deforestation, but with contributions from degradation of agricultural, pasture, and forestlands) and became a net sink sometime around the middle of the last century (although carbon losses from land use change continue at high levels) (*high certainty*). Factors contributing to the growth of the role of ecosystems in carbon sequestration include afforestation, reforestation, and forest management in North America, Europe, China, and other regions; changed agriculture practices; and the fertilizing effects of nitrogen deposition and increasing atmospheric CO_2 (*high certainty*) (C13.ES).

■ *Nitrogen Cycle:* The total amount of reactive, or biologically available, nitrogen created by human activities increased ninefold between 1890 and 1990, with most of that increase taking place in the second half of the century in association with increased use of fertilizers (S7.3.2). (See Figures 1.5 and 1.6.) A recent study of global human contributions to reactive nitrogen flows projected that flows will increase from approximately 165 teragrams of

reactive nitrogen in 1999 to 270 teragrams in 2050, an increase of 64% (R9 Fig 9.1). More than half of all the synthetic nitrogen fertilizer (which was first produced in 1913) ever used on the planet has been used since 1985 (R9.2). Human activities have now roughly doubled the rate of creation of reactive nitrogen on the land surfaces of Earth (R9.2). The flux of reactive nitrogen to the oceans increased by nearly 80% from 1860 to 1990, from roughly 27 teragrams of nitrogen per year to 48 teragrams in 1990 (R9). (This change is not uniform over Earth, however, and while some regions such as Labrador and Hudson's Bay in Canada have seen little if any change, the fluxes from more developed regions such as the northeastern United States, the watersheds of the North Sea in Europe, and the Yellow River basin in China have increased ten- to fifteenfold.)

■ *Phosphorus Cycle:* The use of phosphorus fertilizers and the rate of phosphorus accumulation in agricultural soils increased nearly threefold between 1960 and 1990, although the rate has declined somewhat since that time (S7 Fig 7.18). The current flux of phosphorus to the oceans is now triple that of background rates (approximately 22 teragrams of phosphorus per year versus the natural flux of 8 teragrams) (R9.2)

Most of the reactive nitrogen produced by humans comes from manufacturing nitrogen for synthetic fertilizer and industrial use. Reactive nitrogen is also created as a by-product of fossil fuel combustion and by some (nitrogen-fixing) crops and trees in agroecosystems. The range of the natural rate of bacterial nitrogen fixation in natural terrestrial ecosystems (excluding fixation in agroecosystems) is shown for comparison. Human activity now produces approximately as much reactive nitrogen as natural processes do on the continents. (Note: The 2050 projection is included in the original study and is not based on MA Scenarios.)

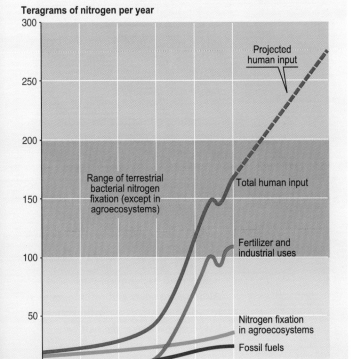

Teragrams of nitrogen per year

Source: Millennium Ecosystem Assessment

Species

A change in an ecosystem necessarily affects the species in the system, and changes in species affect ecosystem processes.

The distribution of species on Earth is becoming more homogenous. By homogenous, we mean that the differences between the set of species at one location on the planet and the set at another location are, on average, diminishing. The natural process of evolution, and particularly the combination of natural barriers to migration and local adaptation of species, led to significant differences in the types of species in ecosystems in different regions. But these regional differences in the planet's biota are now being diminished.

Two factors are responsible for this trend. First, the extinction of species or the loss of populations results in the loss of the presence of species that had been unique to particular regions. Second, the rate of invasion or introduction of species into new ranges is already high and continues to accelerate apace with growing trade and faster transportation. (See Figure 1.7.) For example, a high proportion of the roughly 100 non-native species in the Baltic Sea are native to the North American Great Lakes, and 75% of the recent arrivals of about 170 non-native species in the Great Lakes are native to the Baltic Sea (S10.5). When species decline or go extinct as a result of human activities, they are replaced by a much smaller number of expanding species that thrive in human-altered environments. One effect is that in some regions where diversity has been low, the biotic diversity may actually increase—a result of invasions of non-native forms. (This is true in continental areas such as the Netherlands as well as on oceanic islands.)

Across a range of taxonomic groups, either the population size or range or both of the majority of species is currently declining. Studies of amphibians globally, African mammals, birds in agricultural lands, British butterflies, Caribbean corals, and fishery species show the majority of species to be declining in range or number. Exceptions include species that have been protected in reserves, that have had their particular threats (such as overexploitation) eliminated, or that tend to thrive in landscapes that have been modified by human activity (C4.ES).

Between 10% and 30% of mammal, bird, and amphibian species are currently threatened with extinction (medium to high certainty), based on IUCN–World Conservation Union criteria for threats of extinction. As of 2004, comprehensive assessments of every species within major taxonomic groups have been completed for only three groups of animals (mammals, birds, and amphibians) and two plant groups (conifers and cycads, a group of evergreen palm-like plants). Specialists on these groups have categorized species as "threatened with extinction" if they meet a set of quantitative criteria involving their population size, the size of area in which they are found, and trends in population size or area. (Under the widely used IUCN criteria for extinction, the vast majority of species categorized as "threatened with extinction" have approximately a 10% chance of going extinct within 100 years, although some long-lived species will persist much longer even though their small population size and lack of recruitment means that they have a very high likelihood of extinction.) Twelve percent of bird species, 23% of mammals, and 25% of conifers are currently threatened with extinction; 32% of amphibians are threatened with extinction, but information is more limited and this may be an underestimate. Higher levels of threat have been found in the cycads, where 52% are threatened (C4.ES). In general, freshwater habitats tend to have the highest proportion of threatened species (C4.5.2).

Atmospheric deposition currently accounts for roughly 12% of the reactive nitrogen entering terrestrial and coastal marine ecosystems globally, although in some regions, atmospheric deposition accounts for a higher percentage (about 33% in the United States). (Note: the projection was included in the original study and is not based on MA scenarios.)

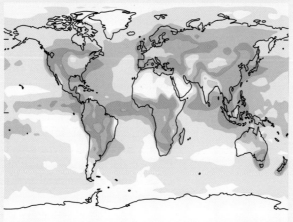

1860

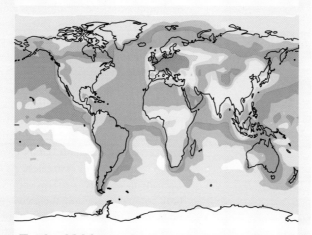

Early 1990s

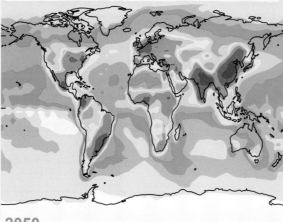

2050

mg nitrogen per sq. meter per year

5   25   50   100   250   500   750   1 000 2 000 5 000

Source: Galloway et al. 2004

Over the past few hundred years, humans have increased the species extinction rate by as much as 1,000 times background rates typical over the planet's history *(medium certainty)* (C4.ES, C4.4.2.). (See Figure 1.8.) Extinction is a natural part of Earth's history. Most estimates of the total number of species today lie between 5 million and 30 million, although the overall total could be higher than 30 million if poorly known groups such as deep-sea organisms, fungi, and microorganisms including parasites have more species than currently estimated. Species present today only represent 2–4% of all species that have ever lived. The fossil record appears to be punctuated by five major mass extinctions, the most recent of which occurred 65 million years ago.

The average rate of extinction found for marine and mammal fossil species (excluding extinctions that occurred in the five major mass extinctions) is approximately 0.1–1 extinctions per million species per year. There are approximately 100 documented extinctions of birds, mammal, and amphibians over the past 100 years, a rate 50–500 times higher than background rates. Including possibly extinct species, the rate is more than 1,000 times higher than background rates. Although the data and techniques used to estimate current extinction rates have improved over the past two decades, significant uncertainty still exists in measuring current rates of extinction because the extent of extinctions of undescribed taxa is unknown, the status of many described species is poorly known, it is difficult to document the final disappearance of very rare species, and there are time lags between the impact of a threatening process and the resulting extinction.

Genes

Genetic diversity has declined globally, particularly among cultivated species. The extinction of species and loss of unique populations has resulted in the loss of unique genetic diversity contained by those species and populations. For wild species, there are few data on the actual changes in the magnitude and distribution of genetic diversity (C4.4), although studies have documented declining genetic diversity in wild species that have been heavily exploited. In cultivated systems, since 1960 there has been a fundamental shift in the pattern of intra-species diversity in farmers' fields and farming systems as the crop varieties planted by farmers have shifted from locally adapted and developed populations (landraces) to more widely adapted varieties produced through formal breeding systems (modern varieties). Roughly 80% of wheat area in developing countries and three quarters of the rice area in Asia is planted with modern varieties (C26.2.1). (For other crops, such as maize, sorghum and millet, the proportion of area planted to modern varieties is far smaller.) The on-farm losses of genetic diversity of crops and livestock have been partially offset by the maintenance of genetic diversity in seed banks.

Figure 1.7. Growth in Number of Marine Species Introductions (C11)

Number of new records of established non-native invertebrate and algae species reported in marine waters of North America, shown by date of first record, and number of new records of non-native marine plant species reported on the European coast, by date of first record.

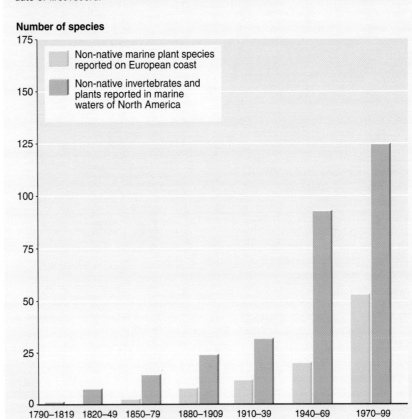

Source: Millennium Ecosystem Assessment

Figure 1.8. Species Extinction Rates (Adapted from C4 Fig 4.22)

"Distant past" refers to average extinction rates as estimated from the fossil record. "Recent past" refers to extinction rates calculated from known extinctions of species (lower estimate) or known extinctions plus "possibly extinct" species (upper bound). A species is considered to be "possibly extinct" if it is believed by experts to be extinct but extensive surveys have not yet been undertaken to confirm its disappearance. "Future" extinctions are model-derived estimates using a variety of techniques, including species-area models, rates at which species are shifting to increasingly more threatened categories, extinction probabilities associated with the IUCN categories of threat, impacts of projected habitat loss on species currently threatened with habitat loss, and correlation of species loss with energy consumption. The time frame and species groups involved differ among the "future" estimates, but in general refer to either future loss of species based on the level of threat that exists today or current and future

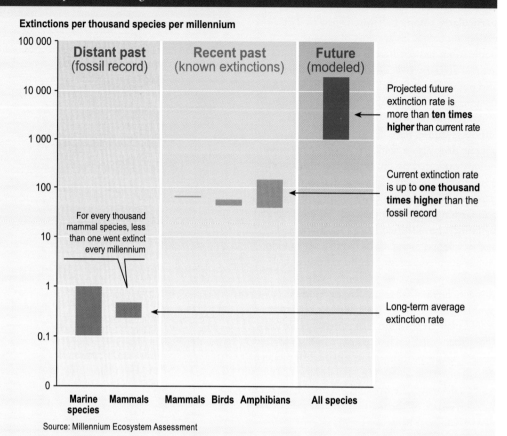

Source: Millennium Ecosystem Assessment

loss of species as a result of habitat changes taking place over the period of roughly 1970 to 2050. Estimates based on the fossil record are *low certainty*; lower-bound estimates for known extinctions are *high certainty* and upper-bound estimates are *medium certainty*; lower-bound estimates for modeled extinctions are *low certainty* and upper-bound estimates are *speculative*. The rate of known extinctions of species in the past century is roughly 50–500 times greater than the extinction rate calculated from the fossil record of 0.1–1 extinctions per 1,000 species per 1,000 years. The rate is up to 1,000 times higher than the background extinction rates if possibly extinct species are included.

2. *How have ecosystem services and their uses changed?*

Ecosystem services are the benefits provided by ecosystems. These include provisioning services such as food, water, timber, fiber, and genetic resources; regulating services such as the regulation of climate, floods, disease, and water quality as well as waste treatment; cultural services such as recreation, aesthetic enjoyment, and spiritual fulfillment; and supporting services such as soil formation, pollination, and nutrient cycling. (See Box 2.1.)

Human use of all ecosystem services is growing rapidly. Approximately 60% (15 out of 24) of the ecosystem services evaluated in this assessment (including 70% of regulating and cultural services) are being degraded or used unsustainably. (See Table 2.1.) Of 24 provisioning, cultural, and regulating ecosystem services for which sufficient information was available, the use of 20 continues to increase. The use of one service, capture fisheries, is now declining as a result of a decline in the quantity of fish, which in turn is due to excessive capture of fish in past decades. Two other services (fuelwood and fiber) show mixed patterns. The use of some types of fiber is increasing and others decreasing; in the case of fuelwood, there is evidence of a recent peak in use.

Humans have enhanced production of three ecosystem services – crops, livestock, and aquaculture – through expansion of the area devoted to their production or through technological inputs. Recently, the service of carbon sequestration has been enhanced globally, due in part to the re-growth of forests in temperate regions, although previously deforestation had been a net source of carbon emissions. Half of provisioning services (6 of 11) and nearly 70% (9 of 13) of regulating and cultural services are being degraded or used unsustainably.

■ *Provisioning Services:* **The quantity of provisioning ecosystem services such as food, water, and timber used by humans increased rapidly, often more rapidly than population growth although generally slower than economic growth, during the second half of the twentieth century. And it continues to grow. In a number of cases, provisioning services are being used at unsustainable rates.** The growing human use has been made possible by a combination of substantial increases in the absolute amount of some services produced by ecosystems and an increase in the fraction used by humans. World population doubled between 1960 and 2000, from 3 billion to 6 billion people, and the global economy increased more than sixfold. During this time, food production increased by roughly two-and-a-half times (a 160% increase in food production between 1961 and 2003), water use doubled, wood harvests for pulp and paper tripled, and timber production increased by nearly 60% (C9.ES, C9.2.2, S7, C7.2.3, C8.1). (Food production increased fourfold in developing countries over this period.)

The sustainability of the use of provisioning services differs in different locations. However, the use of several provisioning services is unsustainable even in the global aggregate. The current level of use of capture fisheries (marine and freshwater) is not sustainable, and many fisheries have already collapsed. (See Figure 2.1.)

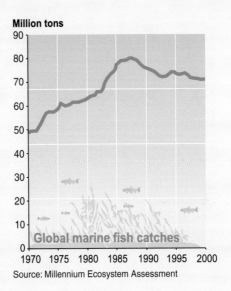

Figure 2.1. Estimated Global Marine Fish Catch, 1950–2001 (C18 Fig 18.3)

In this Figure, the catch reported by governments is in some cases adjusted to correct for likely errors in data.

Million tons

Global marine fish catches

Source: Millennium Ecosystem Assessment

Currently, one quarter of important commercial fish stocks are overexploited or significantly depleted (*high certainty*) (C8.2.2). From 5% to possibly 25% of global freshwater use exceeds long-term accessible supplies and is maintained only through engineered water transfers or the overdraft of groundwater supplies (*low to medium certainty*) (C7.ES). Between 15% and 35% of irrigation withdrawals exceed supply rates and are therefore unsustainable (*low to medium certainty*) (C7.2.2). Current agricultural practices are also unsustainable in some regions due to their reliance on unsustainable sources of water, harmful impacts caused by excessive nutrient or pesticide use, salinization, nutrient depletion, and rates of soil loss that exceed rates of soil formation.

■ *Regulating Services:* **Humans have substantially altered regulating services such as disease and climate regulation by modifying the ecosystem providing the service and, in the case of waste processing services, by exceeding the capabilities of ecosystems to provide the service.** Most changes to regulating services are inadvertent results of actions taken to enhance the supply of provisioning services. Humans have substantially modified the climate regulation service of ecosystems—first through land use changes that contributed to increases in the amount of carbon dioxide and other greenhouse gases such as methane and nitrous oxide in the atmosphere and more recently by increasing the sequestration of carbon dioxide (although ecosystems remain a net source of methane and nitrous oxide). Modifications of

(continued on page 46)

Box 2.1. Ecosystem Services

Ecosystem services are the benefits people obtain from ecosystems. These include provisioning, regulating, and cultural services that directly affect people and the supporting services needed to maintain other services (CF2). Many of the services listed here are highly interlinked. (Primary production, photosynthesis, nutrient cycling, and water cycling, for example, all involve different aspects of the same biological processes.)

Provisioning Services

These are the products obtained from ecosystems, including:

Food. This includes the vast range of food products derived from plants, animals, and microbes.

Fiber. Materials included here are wood, jute, cotton, hemp, silk, and wool.

Fuel. Wood, dung, and other biological materials serve as sources of energy.

Genetic resources. This includes the genes and genetic information used for animal and plant breeding and biotechnology.

Biochemicals, natural medicines, and pharmaceuticals. Many medicines, biocides, food additives such as alginates, and biological materials are derived from ecosystems.

Ornamental resources. Animal and plant products, such as skins, shells, and flowers, are used as ornaments, and whole plants are used for landscaping and ornaments.

Fresh water. People obtain fresh water from ecosystems and thus the supply of fresh water can be considered a provisioning service. Fresh water in rivers is also a source of energy. Because water is required for other life to exist, however, it could also be considered a supporting service.

Regulating Services

These are the benefits obtained from the regulation of ecosystem processes, including:

Air quality regulation. Ecosystems both contribute chemicals to and extract chemicals from the atmosphere, influencing many aspects of air quality.

Climate regulation. Ecosystems influence climate both locally and globally. At a local scale, for example, changes in land cover can affect both temperature and precipitation. At the global scale, ecosystems play an important role in climate by either sequestering or emitting greenhouse gases.

Water regulation. The timing and magnitude of runoff, flooding, and aquifer recharge can be strongly influenced by changes in land cover, including, in particular, alterations that change the water storage potential of the system, such as the conversion of wetlands or the replacement of forests with croplands or croplands with urban areas.

Erosion regulation. Vegetative cover plays an important role in soil retention and the prevention of landslides.

Water purification and waste treatment. Ecosystems can be a source of impurities (for instance, in fresh water) but also can help filter out and decompose organic wastes introduced into inland waters and coastal and marine ecosystems and can assimilate and detoxify compounds through soil and subsoil processes.

Disease regulation. Changes in ecosystems can directly change the abundance of human pathogens, such as cholera, and can alter the abundance of disease vectors, such as mosquitoes.

Pest regulation. Ecosystem changes affect the prevalence of crop and livestock pests and diseases.

Pollination. Ecosystem changes affect the distribution, abundance, and effectiveness of pollinators.

Natural hazard regulation. The presence of coastal ecosystems such as mangroves and coral reefs can reduce the damage caused by hurricanes or large waves.

Cultural Services

These are the nonmaterial benefits people obtain from ecosystems through spiritual enrichment, cognitive development, reflection, recreation, and aesthetic experiences, including:

Cultural diversity. The diversity of ecosystems is one factor influencing the diversity of cultures.

Spiritual and religious values. Many religions attach spiritual and religious values to ecosystems or their components.

Knowledge systems (traditional and formal). Ecosystems influence the types of knowledge systems developed by different cultures.

Educational values. Ecosystems and their components and processes provide the basis for both formal and informal education in many societies.

Inspiration. Ecosystems provide a rich source of inspiration for art, folklore, national symbols, architecture, and advertising.

Aesthetic values. Many people find beauty or aesthetic value in various aspects of ecosystems, as reflected in the support for parks, scenic drives, and the selection of housing locations.

Social relations. Ecosystems influence the types of social relations that are established in particular cultures. Fishing societies, for example, differ in many respects in their social relations from nomadic herding or agricultural societies.

Sense of place. Many people value the "sense of place" that is associated with recognized features of their environment, including aspects of the ecosystem.

Cultural heritage values. Many societies place high value on the maintenance of either historically important landscapes ("cultural landscapes") or culturally significant species.

Recreation and ecotourism. People often choose where to spend their leisure time based in part on the characteristics of the natural or cultivated landscapes in a particular area.

Supporting Services

Supporting services are those that are necessary for the production of all other ecosystem services. They differ from provisioning, regulating, and cultural services in that their impacts on people are often indirect or occur over a very long time, whereas changes in the other categories have relatively direct and short-term impacts on people. (Some services, like erosion regulation, can be categorized as both a supporting and a regulating service, depending on the time scale and immediacy of their impact on people.) These services include:

Soil Formation. Because many provisioning services depend on soil fertility, the rate of soil formation influences human well-being in many ways.

Photosynthesis. Photosynthesis produces oxygen necessary for most living organisms.

Primary production. The assimilation or accumulation of energy and nutrients by organisms.

Nutrient cycling. Approximately 20 nutrients essential for life, including nitrogen and phosphorus, cycle through ecosystems and are maintained at different concentrations in different parts of ecosystems.

Water cycling. Water cycles through ecosystems and is essential for living organisms.

Service	Sub-category	Human Use[a]	Enhanced or Degraded[b]	Notes	MA Chapter
Provisioning Services					
Food	Crops	▲	▲	Food provision has grown faster than overall population growth. Primary source of growth from increase in production per unit area but also significant expansion in cropland. Still persistent areas of low productivity and more rapid area expansion, e.g., sub-Saharan Africa and parts of Latin America.	C8.2
	Livestock	▲	▲	Significant increase in area devoted to livestock in some regions, but major source of growth has been more intensive, confined production of chicken, pigs, and cattle.	C8.2
	Capture fisheries	▼	▼	Marine fish harvest increased until the late 1980s and has been declining since then. Currently, one quarter of marine fish stocks are overexploited or significantly depleted. Freshwater capture fisheries have also declined. Human use of capture fisheries as declined because of the reduced supply, not because of reduced demand.	C18 C8.2.2 C19
	Aquaculture	▲	▲	Aquaculture has become a globally significant source of food in the last 50 years and, in 2000, contributed 27% of total fish production. Use of fish feed for carnivorous aquaculture species places an additional burden on capture fisheries.	C8 Table 8.4
	Wild plant and animal products	NA	▼	Provision of these food sources is generally declining as natural habitats worldwide are under increasing pressure and as wild populations are exploited for food, particularly by the poor, at unsustainable levels.	C8.3.1
Fiber	Timber	▲	+/−	Global timber production has increased by 60% in the last four decades. Plantations provide an increasing volume of harvested roundwood, amounting to 35% of the global harvest in 2000. Roughly 40% of forest area has been lost during the industrial era, and forests continue to be lost in many regions (thus the service is degraded in those regions), although forest is now recovering in some temperate countries and thus this service has been enhanced (from this lower baseline) in these regions in recent decades.	C9.ES C21.1
	Cotton, hemp, silk	+/−	+/−	Cotton and silk production have doubled and tripled respectively in the last four decades. Production of other agricultural fibers has declined.	C9.ES
	Wood fuel	+/−	▼	Global consumption of fuelwood appears to have peaked in the 1990s and is now believed to be slowly declining but remains the dominant source of domestic fuel in some regions.	C9.ES
Genetic resources		▲	▼	Traditional crop breeding has relied on a relatively narrow range of germplasm for the major crop species, although molecular genetics and biotechnology provide new tools to quantify and expand genetic diversity in these crops. Use of genetic resources also is growing in connection with new industries based on biotechnology. Genetic resources have been lost through the loss of traditional cultivars of crop species (due in part to the adoption of modern farming practices and varieties) and through species extinctions.	C26.2.1

(continued on page 42)

Service	Sub-category	Human Use[a]	Enhanced or Degraded[b]	Notes	MA Chapter
Biochemicals, natural medicines, and pharmaceuticals		▲	▼	Demand for biochemicals and new pharmaceuticals is growing, but new synthetic technologies compete with natural products to meet the demand. For many other natural products (cosmetics, personal care, bioremediation, biomonitoring, ecological restoration), use is growing. Species extinction and overharvesting of medicinal plants is diminishing the availability of these resources.	C10
Ornamental resources		NA	NA		
Fresh water		▲	▼	Human modification of ecosystems (e.g., reservoir creation) has stabilized a substantial fraction of continental river flow, making more fresh water available to people but in dry regions reducing river flows through open water evaporation and support to irrigation that also loses substantial quantities of water. Watershed management and vegetation changes have also had an impact on seasonal river flows. From 5% to possibly 25% of global freshwater use exceeds long-term accessible supplies and requires supplies either through engineered water transfers or overdraft of groundwater supplies. Between 15% and 35% of irrigation withdrawals exceed supply rates. Fresh water flowing in rivers also provides a service in the form of energy that is exploited through hydropower. The construction of dams has not changed the amount of energy, but it has made the energy more available to people. The installed hydroelectric capacity doubled between 1960 and 2000. Pollution and biodiversity loss are defining features of modern inland water systems in many populated parts of the world.	C7

Regulating Services

Service	Sub-category	Human Use[a]	Enhanced or Degraded[b]	Notes	MA Chapter
Air quality regulation		▲	▼	The ability of the atmosphere to cleanse itself of pollutants has declined slightly since preindustrial times but likely not by more than 10%. The net contribution of ecosystems to this change is not known. Ecosystems are also a sink for tropospheric ozone, ammonia, NO_x, SO_2, particulates, and CH_4, but changes in these sinks were not assessed.	C13.ES
Climate regulation	Global	▲	▲	Terrestrial ecosystems were on average a net source of CO_2 during the nineteenth and early twentieth century and became a net sink sometime around the middle of the last century. The biophysical effect of historical land cover changes (1750 to present) is net cooling on a global scale due to increased albedo, partially offsetting the warming effect of associated carbon emissions from land cover change over much of that period.	C13.ES
	Regional and local	▲	▼	Changes in land cover have affected regional and local climates both positively and negatively, but there is a preponderance of negative impacts. For example, tropical deforestation and desertification have tended to reduce local rainfall.	C13.3 C11.3
Water regulation		▲	+/–	The effect of ecosystem change on the timing and magnitude of runoff, flooding, and aquifer recharge depends on the ecosystem involved and on the specific modifications made to the ecosystem.	C7.4.4

Service	Sub-category	Human Use[a]	Enhanced or Degraded[b]	Notes	MA Chapter
Erosion regulation		▲	▼	Land use and crop/soil management practices have exacerbated soil degradation and erosion, although appropriate soil conservation practices that reduce erosion, such as minimum tillage, are increasingly being adopted by farmers in North America and Latin America.	C26
Water purification and waste treatment		▲	▼	Globally, water quality is declining, although in most industrial countries pathogen and organic pollution of surface waters has decreased over the last 20 years. Nitrate concentration has grown rapidly in the last 30 years. The capacity of ecosystems to purify such wastes is limited, as evidenced by widespread reports of inland waterway pollution. Loss of wetlands has further decreased the ability of ecosystems to filter and decompose wastes.	C7.2.5 C19
Disease regulation		▲	+/−	Ecosystem modifications associated with development have often increased the local incidence of infectious diseases, although major changes in habitats can both increase or decrease the risk of particular infectious diseases.	C14
Pest regulation		▲	▼	In many agricultural areas, pest control provided by natural enemies has been replaced by the use of pesticides. Such pesticide use has itself degraded the capacity of agroecosystems to provide pest control. In other systems, pest control provided by natural enemies is being used and enhanced through integrated pest management. Crops containing pest-resistant genes can also reduce the need for application of toxic synthetic pesticides.	C11.3
Pollination		▲	▼[c]	There is *established but incomplete* evidence of a global decline in the abundance of pollinators. Pollinator declines have been reported in at least one region or country on every continent except Antarctica, which has no pollinators. Declines in abundance of pollinators have rarely resulted in complete failure to produce seed or fruit, but more frequently resulted in fewer seeds or in fruit of reduced viability or quantity. Losses in populations of specialized pollinators have directly affected the reproductive ability of some rare plants.	C11 Box 11.2
Natural hazard regulation		▲	▼	People are increasingly occupying regions and localities that are exposed to extreme events, thereby exacerbating human vulnerability to natural hazards. This trend, along with the decline in the capacity of ecosystems to buffer from extreme events, has led to continuing high loss of life globally and rapidly rising economic losses from natural disasters.	C16 C19
Cultural Services					
Cultural diversity		NA	NA		

(continued on page 44)

Service	Sub-category	Human Use[a]	Enhanced or Degraded[b]	Notes	MA Chapter
Cultural Services *(continued)*					
Spiritual and religious values		▲	▼	There has been a decline in the numbers of sacred groves and other such protected areas. The loss of particular ecosystem attributes (sacred species or sacred forests), combined with social and economic changes, can sometimes weaken the spiritual benefits people obtain from ecosystems. On the other hand, under some circumstances (e.g., where ecosystem attributes are causing significant threats to people), the loss of some attributes may enhance spiritual appreciation for what remains.	C17.2.3
Knowledge systems		NA	NA		
Educational values		NA	NA		
Inspiration		NA	NA		
Aesthetic values		▲	▼	The demand for aesthetically pleasing natural landscapes has increased in accordance with increased urbanization. There has been a decline in quantity and quality of areas to meet this demand. A reduction in the availability of and access to natural areas for urban residents may have important detrimental effects on public health and economies.	C17.2.5
Social relations		NA	NA		
Sense of place		NA	NA		
Cultural heritage values		NA	NA		
Recreation and ecotourism		▲	+/–	The demand for recreational use of landscapes is increasing, and areas are increasingly being managed to cater for this use, to reflect changing cultural values and perceptions. However, many naturally occurring features of the landscape (e.g., coral reefs) have been degraded as resources for recreation.	C17.2.6 C19
Supporting Services					
Soil formation		†	†		
Photosynthesis		†	†		
Primary production		†	†	Several global MA systems, including dryland, forest, and cultivated systems, show a trend of NPP increase for the period 1981 to 2000. However, high seasonal and inter-annual variations associated with climate variability occur within this trend on the global scale	C22.2.1

Service	Sub-category	Human Use[a]	Enhanced or Degraded[b]	Notes	MA Chapter
Supporting Services *(continued)*					
Nutrient cycling		†	†	There have been large-scale changes in nutrient cycles in recent decades, mainly due to additional inputs from fertilizers, livestock waste, human wastes, and biomass burning. Inland water and coastal systems have been increasingly affected by eutrophication due to transfer of nutrients from terrestrial to aquatic systems as biological buffers that limit these transfers have been significantly impaired.	C12 S7
Water cycling		†	†	Humans have made major changes to water cycles through structural changes to rivers, extraction of water from rivers, and, more recently, climate change.	C7

[a] For provisioning services, human use increases if the human consumption of the service increases (e.g., greater food consumption); for regulating and cultural services, human use increases if the number of people affected by the service increases. The time frame is in general the past 50 years, although if the trend has changed within that time frame, the indicator shows the most recent trend.

[b] For provisioning services, we define enhancement to mean increased production of the service through changes in area over which the service is provided (e.g., spread of agriculture) or increased production per unit area. We judge the production to be degraded if the current use exceeds sustainable levels. For regulating and supporting services, enhancement refers to a change in the service that leads to greater benefits for people (e.g., the service of disease regulation could be improved by eradication of a vector known to transmit a disease to people). Degradation of a regulating and supporting services means a reduction in the benefits obtained from the service, either through a change in the service (e.g., mangrove loss reducing the storm protection benefits of an ecosystem) or through human pressures on the service exceeding its limits (e.g., excessive pollution exceeding the capability of ecosystems to maintain water quality). For cultural services, degradation refers to a change in the ecosystem features that decreases the cultural (recreational, aesthetic, spiritual, etc.) benefits provided by the ecosystem. The time frame is in general the past 50 years, although if the trend has changed within that time frame the indicator shows the most recent trend.

[c] *Low to medium certainty.* All other trends are *medium to high certainty.*

Legend:

▲ = Increasing (for human use column) or enhanced (for enhanced or degraded column)

▼ = Decreasing (for human use column) or degraded (for enhanced or degraded column)

+/– = Mixed (trend increases and decreases over past 50 years or some components/regions increase while others decrease)

NA = Not assessed within the MA. In some cases, the service was not addressed at all in the MA (such as ornamental resources), while in other cases the service was included but the information and data available did not allow an assessment of the pattern of human use of the service or the status of the service.

† = The categories of "human use" and "enhanced or degraded" do not apply for supporting services since, by definition, these services are not directly used by people. (Their costs or benefits would be double-counted if the indirect effects were included.) Changes in supporting services influence the supply of provisioning, cultural, or regulating services that are then used by people and may be enhanced or degraded.

ecosystems have altered patterns of disease by increasing or decreasing habitat for certain diseases or their vectors (such as dams and irrigation canals that provide habitat for schistosomiasis) or by bringing human populations into closer contact with various disease organisms. Changes to ecosystems have contributed to a significant rise in the number of floods and major wildfires on all continents since the 1940s. Ecosystems serve an important role in detoxifying wastes introduced into the environment, but there are intrinsic limits to that waste processing capability. For example, aquatic ecosystems "cleanse" on average 80% of their global incident nitrogen loading, but this intrinsic self-purification capacity varies widely and is being reduced by the loss of wetlands (C7.2.5).

■ *Cultural Services:* Although the use of cultural services has continued to grow, the capability of ecosystems to provide cultural benefits has been significantly diminished in the past century (C17). Human cultures are strongly influenced by ecosystems, and ecosystem change can have a significant impact on cultural identity and social stability. Human cultures, knowledge systems, religions, heritage values, social interactions, and the linked amenity services (such as aesthetic enjoyment, recreation, artistic and spiritual fulfillment, and intellectual development) have always been influenced and shaped by the nature of the ecosystem and ecosystem conditions. Many of these benefits are being degraded, either through changes to ecosystems (a recent rapid decline in the numbers of sacred groves and other such protected areas, for example) or through societal changes (such as the loss of languages or of traditional knowledge) that reduce people's recognition or appreciation of those cultural benefits. Rapid loss of culturally valued ecosystems and landscapes can contribute to social disruptions and societal marginalization. And there has been a decline in the quantity and quality of aesthetically pleasing natural landscapes.

Global gains in the supply of food, water, timber, and other provisioning services were often achieved in the past century despite local resource depletion and local restrictions on resource use by shifting production and harvest to new underexploited regions, sometimes considerable distances away. These options are diminishing. This trend is most distinct in the case of marine fisheries. As individual stocks have been depleted, fishing pressure has shifted to less exploited stocks (C18.2.1). Industrial fishing fleets have also shifted to fishing further offshore and in deeper water to meet global demand (C18.ES). (See Figure 2.2.) A variety of drivers related to market demand, supply, and government policies have influenced patterns of timber harvest. For example, international trade in forest products increases when a nation's forests no longer can meet demand or when policies have been established to restrict or ban timber harvest.

Although human demand for ecosystem services continues to grow in the aggregate, the demand for particular services in specific regions is declining as substitutes are developed. For example, kerosene, electricity, and other energy sources are increasingly being substituted for fuelwood (still the primary source of energy for heating and cooking for some 2.6 billion people) (C9.ES). The substitution of a variety of other materials

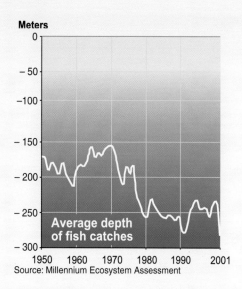

Figure 2.2. TREND IN MEAN DEPTH OF CATCH SINCE 1950

Fisheries catches increasingly originate from deep areas. (Data from C18 Fig 18.5)

for wood (such as vinyl, plastics, and metal) has contributed to relatively slow growth in global timber consumption in recent years (C9.2.1). While the use of substitutes can reduce pressure on specific ecosystem services, this may not always have positive net environmental benefits. Substitution of fuelwood by fossil fuels, for example, reduces pressure on forests and lowers indoor air pollution, but it may increase net greenhouse gas emissions. Substitutes are also often costlier to provide than the original ecosystem services.

Both the supply and the resilience of ecosystem services are affected by changes in biodiversity. Biodiversity is the variability among living organisms and the ecological complexes of which they are part. When a species is lost from a particular location (even if it does not go extinct globally) or introduced to a new location, the various ecosystem services associated with that species are changed. More generally, when a habitat is converted, an array of ecosystem services associated with the species present in that location is changed, often with direct and immediate

impacts on people (S10). Changes in biodiversity also have numerous indirect impacts on ecosystem services over longer time periods, including influencing the capacity of ecosystems to adjust to changing environments (*medium certainty*), causing disproportionately large and sometimes irreversible changes in ecosystem processes, influencing the potential for infectious disease transmission, and, in agricultural systems, influencing the risk of crop failure in a variable environment and altering the potential impacts of pests and pathogens (*medium to high certainty*) (C11.ES, C14.ES).

The modification of an ecosystem to alter one ecosystem service (to increase food or timber production, for instance) generally results in changes to other ecosystem services as well (CWG, SG7). Trade-offs among ecosystem services are commonplace. (See Table 2.2.) For example, actions to increase food production often involve one or more of the following: increased water use, degraded water quality, reduced biodiversity, reduced forest cover, loss of forest products, or release of greenhouse gases. Frequent cultivation, irrigated rice production, livestock production, and burning of cleared areas and crop residues now release 1,600±800 million tons of carbon per year in CO_2 (C26. ES). Cultivation, irrigated rice production, and livestock production release between 106 million and 201 million tons of carbon per year in methane (C13 Table 13.1). About 70% of anthropogenic nitrous oxide gas emissions are attributable to agriculture, mostly from land conversion and nitrogen fertilizer use (C26. ES). Similarly, the conversion of forest to agriculture can significantly change flood frequency and magnitude, although the amount and direction of this impact is highly dependent on the characteristics of the local ecosystem and the nature of the land cover change (C21.5.2).

Many trade-offs associated with ecosystem services are expressed in areas remote from the site of degradation. For example, conversion of forests to agriculture can affect water quality and flood frequency downstream of where the ecosystem change occurred. And increased application of nitrogen fertilizers to croplands can have negative impacts on coastal water quality. These trade-offs are rarely taken fully into account in decision-making, partly due to the sectoral nature of planning and partly because some of the effects are also displaced in time (such as long-term climate impacts).

The net benefits gained through actions to increase the productivity or harvest of ecosystem services have been less than initially believed after taking into account negative trade-offs. The benefits of resource management actions have traditionally been evaluated only from the standpoint of the service targeted by the management intervention. However, management interventions to increase any particular service almost always result in costs to other services. Negative trade-offs are commonly found between individual provisioning services and between provisioning services and the combined regulating, cultural, and supporting services and biodiversity. Taking the costs of these negative trade-offs into account reduces the apparent benefits of the various management interventions. For example:

■ Expansion of commercial shrimp farming has had serious impacts on ecosystems, including loss of vegetation, deterioration of water quality, decline of capture fisheries, and loss of biodiversity (R6, C19).

■ Expansion of livestock production around the world has often led to overgrazing and dryland degradation, rangeland fragmentation, loss of wildlife habitat, dust formation, bush encroachment, deforestation, nutrient overload through disposal of manure, and greenhouse gas emissions (R6.ES).

■ Poorly designed and executed agricultural policies led to an irreversible change in the Aral Sea ecosystem. By 1998, the Aral Sea had lost more than 60% of its area and approximately 80% of its volume, and ecosystem-related problems in the region now include excessive salt content of major rivers, contamination of agricultural products with agrochemicals, high levels of turbidity in major water sources, high levels of pesticides and phenols in surface waters, loss of soil fertility, extinctions of species, and destruction of commercial fisheries (R6 Box 6.9).

■ Forested riparian wetlands adjacent to the Mississippi river in the United States had the capacity to store about 60 days of river discharge. With the removal of the wetlands through canalization, leveeing, and draining, the remaining wetlands have a storage capacity of less than 12 days discharge, an 80% reduction in flood storage capacity (C16.1.1).

However, positive synergies can be achieved as well when actions to conserve or enhance a particular component of an ecosystem or its services benefit other services or stakeholders. Agroforestry can meet human needs for food and fuel, restore soils, and contribute to biodiversity conservation. Intercropping can increase yields, increase biocontrol, reduce soil erosion, and reduce weed invasion in fields. Urban parks and other urban green spaces provide spiritual, aesthetic, educational, and recreational benefits as well as such services such as water purification, wildlife habitat, waste management, and carbon sequestration. Protection of natural forests for biodiversity conservation can also reduce carbon emissions and protect water supplies. Protection of wetlands can contribute to flood control and also help to remove pollutants such as phosphorus and nitrogen from the water. For example, it is estimated that the nitrogen load from the heavily polluted Illinois River basin to the Mississippi River could be cut in half by converting 7% of the basin back to wetlands (R9.4.5). Positive synergies often exist among regulating, cultural, and supporting services and with biodiversity conservation.

Table 2.2. INDICATIVE ECOSYSTEM SERVICE TRADE-OFFS

The nature and direction of trade-offs among ecosystem services depends significantly on the specific management practices used to change the target service and on the ecosystem involved. This table summarizes common directions of trade-offs encountered across ecosystem services, although the magnitude (or even direction) of the trade-off may differ from case to case.

Management Practice	Provisioning Services			Regulating Services			Cultural Services	Supporting Services	Notes
	Food Production	Water Availability and Quality	Fiber Production	Carbon Sequestration	Disease Reduction	Flood Control	Ecotourism Potential	N Regulation (Avoidance of Eutrophication)	
Increased food production through intensification of agriculture	Intervention target	−	o	−	+/−	o	o	−	Agricultural ecosystems reduce exposure to certain diseases but increase the risk of other diseases
Increased food production through expansion of agriculture	Intervention target	−	−	−	+/−	−	−	−	
Increased wild fish catch	Intervention target	NA	NA	NA	NA	NA	+/−	+/−	Increased fish catch can increase ecotourism opportunities (e.g., increased sport fishing opportunities) or decrease them if the levels are unsustainable or if the increased catch reduces populations of predators that attract tourists (e.g., killer whales, seals, sea lions).
Damming rivers to increase water availability	+	Intervention target	−	+/−	−	+/−	+/−	−	River modification can reduce flood frequency but increase the risk and magnitude of catastrophic floods. Reservoirs provide some recreational opportunities but those associated with the original river are lost.
Increased timber harvest	−	+/−	Intervention target	−	+/−	+/−	−	o	Timber harvest generally reduces availability of wild sources of food.
Draining or filling wetlands to reduce malaria risk	+	−	o	o	Intervention target	−	−	−	Filled wetlands are often used for agriculture. Loss of wetlands results in a loss of water cleansing capability, loss of a source of flood control and ecotourism potential.
Establishing a strictly protected area to maintain biodiversity and provide recreation	−	+	−	+	+/−	+	+	+	Strictly protected areas may result in the loss of a local source of food supply and fiber production. The presence of the protected area safeguards water supplies and water quality, prevents emissions of greenhouse gases that might have resulted from habitat conversion and increases tourism potential.

Legend:
- − = change in the first column has a negative impact on the service
- + = change in the first column has a positive impact on the service
- o = change in the first column is neutral or has no effect on the service
- NA = the category is not applicable

3. *How have ecosystem changes affected human well-being and poverty alleviation?*

Relationships between Ecosystem Services and Human Well-being

Changes in ecosystem services influence all components of human well-being, including the basic material needs for a good life, health, good social relations, security, and freedom of choice and action (CF3). (See Box 3.1.) Humans are fully dependent on Earth's ecosystems and the services that they provide, such as food, clean water, disease regulation, climate regulation, spiritual fulfillment, and aesthetic enjoyment. The relationship between ecosystem services and human well-being is mediated by access to manufactured, human, and social capital. Human well-being depends on ecosystem services but also on the supply and quality of social capital, technology, and institutions. These factors mediate the relationship between ecosystem services and human well-being in ways that remain contested and incompletely understood. The relationship between human well-being and ecosystem services is not linear. When an ecosystem service is abundant relative to the demand, a marginal increase in ecosystem services generally contributes only slightly to human well-being (or may even diminish it). But when the service is relatively scarce, a small decrease can substantially reduce human well-being (S.SDM, SG3.4).

Ecosystem services contribute significantly to global employment and economic activity. The ecosystem service of food production contributes by far the most to economic activity and employment. In 2000, the market value of food production was $981 billion, or roughly 3% of gross world product, but it is a much higher share of GDP within developing countries (C8 Table 8.1). That year, for example, agriculture (including forestry and fishing) represented 24% of total GDP in countries with per capita incomes less than $765 (the low-income developing countries, as defined by the World Bank) (C26.5.1). The agricultural labor force contained 1.3 billion people globally—approximately a fourth (22%) of the world's population and half (46%) of the total labor force—and some 2.6 billion people, more than 40% of the world, lived in agriculturally based households (C26.5.1). Significant differences exist between developing and industrial countries in these patterns. For example, in the United States only 2.4% of the labor force works in agriculture.

Other ecosystem services (or commodities based on ecosystem services) that make significant contributions to national economic activity include timber (around $400 billion), marine fisheries (around $80 billion in 2000), marine aquaculture ($57 billion in 2000), recreational hunting and fishing ($50 billion and $24–37 billion annually respectively in the United States alone), as well as edible forest products, botanical medicines, and medicinal plants (C9.ES, C18.1, C20.ES). And many other industrial products and commodities rely on ecosystem services such as water as inputs.

The degradation of ecosystem services represents a loss of a capital asset (C5.4.1). (See Figure 3.1.) Both renewable resources such as ecosystem services and nonrenewable resources such as mineral deposits, soil nutrients, and fossil fuels are capital assets. Yet traditional national accounts do not include measures of resource depletion or of the degradation of renewable resources. As a result, a country could cut its forests and deplete its fisheries, and this would show only as a positive gain to GDP despite the loss of the capital asset. Moreover, many ecosystem services are available freely to those who use them (fresh water in aquifers, for instance, or the use of the atmosphere as a sink for pollutants), and so again their degradation is not reflected in standard economic measures.

When estimates of the economic losses associated with the depletion of natural assets are factored into measurements of the total wealth of nations, they significantly change the balance sheet of those countries with economies especially dependent on natural resources. For example, countries such as Ecuador, Ethiopia, Kazakhstan, Republic of Congo, Trinidad and Tobago, Uzbekistan, and Venezuela that had positive growth in net savings (reflecting a growth in the net wealth of the country) in 2001 actually experienced a loss in net savings when depletion of natural resources (energy and forests) and estimated damages from carbon emissions (associated with contributions to climate change) were factored into the accounts. In 2001, in 39 countries out of the 122 countries for which sufficient data were available, net national savings (expressed as a percent of gross national income) were reduced by at least 5% when costs associated with the depletion of natural resources (unsustainable forestry, depletion of fossil fuels) and damage from carbon emissions were included.

The degradation of ecosystem services often causes significant harm to human well-being (C5 Box 5.2). The information available to assess the consequences of changes in ecosystem services for human well-being is relatively limited. Many ecosystem services have not been monitored and it is also difficult to estimate the relative influence of changes in ecosystem services in relation to other social, cultural, and economic factors that also affect human well-being. Nevertheless, the following evidence demonstrates that the harmful effects of the degradation of ecosystem services on livelihoods, health, and local and national economies are substantial.

■ *Most resource management decisions are most strongly influenced by ecosystem services entering markets; as a result, the nonmarketed benefits are often lost or degraded.* Many ecosystem services, such as the purification of water, regulation of floods, or provision of

(continued on page 56)

Box 3.1. LINKAGES BETWEEN ECOSYSTEM SERVICES AND HUMAN WELL-BEING

Human well-being has five main components: the basic material needs for a good life, health, good social relations, security, and freedom of choice and action. (See Box Figure A.) This last component is influenced by other constituents of well-being (as well as by other factors including, notably, education) and is also a precondition for achieving other components of well-being, particularly with respect to equity and fairness. Human well-being is a continuum—from extreme deprivation, or poverty, to a high attainment or experience of well-being. Ecosystems underpin human well-being through supporting, provisioning, regulating, and cultural services. Well-being also depends on the supply and quality of human services, technology, and institutions.

Basic Materials for a Good Life
This refers to the ability to have a secure and adequate livelihood, including income and assets, enough food and water at all times, shelter, ability to have energy to keep warm and cool, and access to goods. Changes in provisioning services such as food, water, and fuelwood have very strong impacts on the adequacy of material for a good life. Access to these materials is heavily mediated by socioeconomic circumstances. For the wealthy, local changes in ecosystems may not cause a significant change in their access to necessary material goods, which can be purchased from other locations, sometimes at artificially low

---

Box Figure A. ILLUSTRATION OF LINKAGES BETWEEN ECOSYSTEM SERVICES AND HUMAN WELL-BEING

This figure depicts the strength of linkages between categories of ecosystem services and components of human well-being that are commonly encountered, and includes indications of the extent to which it is possible for socioeconomic factors to mediate the linkage. (For example, if it is possible to purchase a substitute for a degraded ecosystem service, then there is a high potential for mediation.) The strength of the linkages and the potential for mediation differ in different ecosystems and regions. In addition to the influence of ecosystem services on human well-being depicted here, other factors—including other environmental factors as well as economic, social, technological, and cultural factors—influence human well-being, and ecosystems are in turn affected by changes in human well-being.

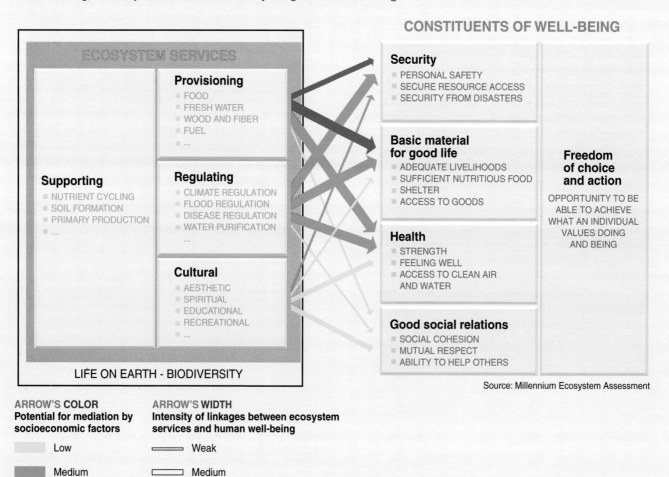

Source: Millennium Ecosystem Assessment

prices if governments provide subsidies (for example, water delivery systems). Changes in regulating services influencing water supply, pollination and food production, and climate have very strong impacts on this element of human well-being. These, too, can be mediated by socioeconomic circumstances, but to a smaller extent. Changes in cultural services have relatively weak linkages to material elements of well-being. Changes in supporting services have a strong influence by virtue of their influence on provisioning and regulating services. The following are some examples of material components of well-being affected by ecosystem change.

■ *Income and Employment:* Increased production of crops, fisheries, and forest products has been associated with significant growth in local and national economies. Changes in the use and management of these services can either increase employment (as, for example, when agriculture spreads to new regions) or decrease it through gains in productivity of labor. In regions where productivity has declined due to land degradation or overharvesting of fisheries, the impacts on local economies and employment can be devastating to the poor or to those who rely on these services for income.

■ *Food:* The growth in food production and farm productivity has more than kept pace with global population growth, resulting in significant downward pressure on the price of foodstuffs. Following significant spikes in the 1970s caused primarily by oil crises, there have been persistent and profound reductions in the price of foodstuffs globally (C8.1). Over the last 40 years, food prices have dropped by around 40% in real terms due to increases in productivity (C26.2.3). It is *well established* that past increases in food production, at progressively lower unit cost, have improved the health and well-being of billions, particularly the most needy, who spend the largest share of their incomes on food (C8.1). Increased production of food and lower prices for food have not been entirely positive. Among industrial countries, and increasingly among developing ones, diet-related risks, mainly associated with overnutrition, in combination with physical inactivity now account for one

third of the burden of disease (R16.1.2). At present, over 1 billion adults are overweight, with at least 300 million considered clinically obese, up from 200 million in 1995 (C8.5.1).

■ *Water Availability:* The modification of rivers and lakes through the construction of dams and diversions has increased the water available for human use in many regions of the world. However, the declining per capita availability of water is having negative impacts on

human well-being. Water scarcity is a globally significant and accelerating condition for roughly 1–2 billion people worldwide, leading to problems with food production, human health, and economic development. Rates of increase in a key water scarcity measure (water use relative to accessible supply) from 1960 to the present averaged nearly 20% per decade globally, with values of 15% to more than 30% per decade for individual continents (C7.ES).

Box Table. SELECTED WATER-RELATED DISEASES

Approximate yearly number of cases, mortality, and disability-adjusted life years. The DALY is a summary measure of population health, calculated on a population scale as the sum of years lost due to premature mortality and of healthy years lost to disability for incident cases of the ill-health condition (C7 Table 7.10).

Disease	Number of Cases	Disability-Adjusted Life Years (thousand DALYs)	Estimated Mortality (thousand)	Relationship to Freshwater Services
Diarrhea	4 billion	62,000 (54,000)[a]	1,800 (1,700)[a]	water contaminated by human feces
Malaria	300–500 million	46,500	1,300	transmitted by Anopheles mosquitoes
Schistosomiasis	200 million	1,700	15	transmitted by aquatic mollusks
Dengue and dengue hemorrhagic fever	50–100 million dengue; 500,000 DHF	616	19	transmitted by Aedes mosquitoes
Onchocerciasis (river blindness)	18 million	484	0	transmitted by black fly
Typhoid and paratyphoid fevers	17 million			contaminated water, food, flooding
Trachoma	150 million, with 6 million blind	2,300	0	lack of basic hygiene
Cholera	140,000–184,000[a]		5–28[b]	water and food contaminated by human feces
Dracunculiasis (Guinea worm disease)	96,000			contaminated water

[a] Diarrhea is a water-related disease, but not all diarrhea is associated with contaminated water. The number in parentheses refers to the diarrhea specifically associated with contaminated water.
[b] The upper part of the range refers specifically to 2001.

(continued on page 52)

Health

By health, we refer to the ability of an individual to feel well and be strong, or in other words to be adequately nourished and free from disease, to have access to adequate and clean drinking water and clean air, and to have the ability to have energy to keep warm and cool. Human health is both a product and a determinant of well-being. Changes in provisioning services such as food, water, medicinal plants, and access to new medicines and changes in regulating services that influence air quality, water quality, disease regulation, and waste treatment also have very strong impacts on health. Changes in cultural services can have strong influences on health,

since they affect spiritual, inspirational, aesthetic, and recreational opportunities, and these in turn affect both physical and emotional states. Changes in supporting services have a strong influence on all of the other categories of services. These benefits are moderately mediated by socioeconomic circumstances. The wealthy can purchase substitutes for some health benefits of ecosystems (such as medicinal plants or water quality), but they are more susceptible to changes affecting air quality. The following are some examples of health components of well-being affected by ecosystem change.

■ *Nutrition:* In 2000, about a quarter of the burden of disease among the poorest

countries was attributable to childhood and maternal undernutrition. Worldwide, undernutrition accounted for nearly 10% of the global burden of disease (R16.1.2).

■ *Water and Sanitation:* The burden of disease from inadequate water, sanitation, and hygiene totals 1.7 million deaths and results in the loss of at least 54 million healthy life years annually. Along with sanitation, water availability and quality are well recognized as important risk factors for infectious diarrhea and other major diseases. (See Box Table.) Some 1.1 billion people lack access to clean drinking water, and more than 2.6 billion lack access to sanitation (C7.ES). (See Box Figures B and C.) Globally, the economic cost of pollution of

Box Figure B. Proportion of Population with Improved Drinking Water Supply in 2002 (C7 Fig 7.13)

Access to improved drinking water is estimated by the percentage of the population using the following drinking water sources: household connection, public standpipe, borehole, protected dug well, protected spring, or rainwater collection.

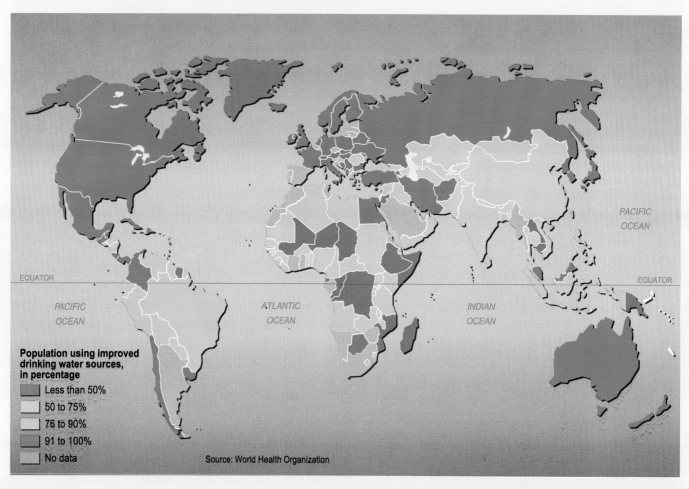

Population using improved drinking water sources, in percentage
- Less than 50%
- 50 to 75%
- 76 to 90%
- 91 to 100%
- No data

Source: World Health Organization

coastal waters is estimated to be $16 billion annually, mainly due to human health impacts (C19.3.1).

■ *Vector-borne Disease:* Actions to reduce vector-borne diseases have resulted in major health gains and helped to relieve important constraints on development in poor regions. Vector-borne diseases cause approximately 1.4 million deaths a year, mainly due to malaria in Africa. These infections are both an effect and a cause of poverty (R12.ES). Prevalence of a number of infectious diseases appears to be growing, and environmental changes such as deforestation, dam construction, road building, agricultural conversion,

and urbanization are contributing factors in many cases (C14.2).

■ *Medicines:* The use of natural products in the pharmaceutical industry has tended to fluctuate widely, with a general decline in pharmaceutical bioprospecting by major companies. Historically, most drugs were obtained from natural products. Even near the end of the twentieth century, approximately 50% of prescription medicines were originally discovered in plants (C10.2). Natural products still are actively used in drug exploration. Medicinal plants continue to play an important role in health care systems in many parts of the world. One MA subglobal assessment in the Mekong wetlands

identified more than 280 medically important plant species, of which 150 are still in regular use (C10.2.2). Medicinal plants have generally declined in availability due to overharvesting and loss of habitats (C10.5.4).

Good Social Relations

Good social relations refer to the presence of social cohesion, mutual respect, and the ability to help others and provide for children. Changes in provisioning and regulating ecosystem services can affect social relations, principally through their more direct impacts on material well-being, health, and security. Changes in cultural services can have a strong

---

Box Figure C. PROPORTION OF POPULATION WITH IMPROVED SANITATION COVERAGE IN 2002 (C7 Fig 7.14)

Access to improved sanitation is estimated by the percentage of the population using the following sanitation facilities: connection to a public sewer, connection to a septic system, pour-flush latrine, simple pit latrine (a portion of pit latrines are also considered unimproved sanitation), and ventilated improved pit latrine.

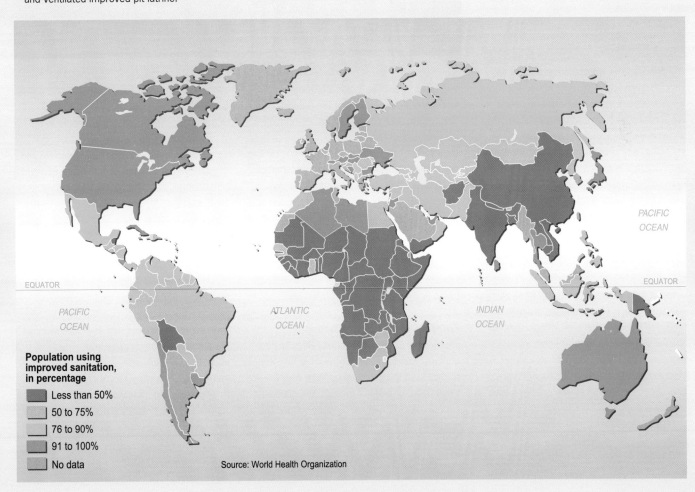

Population using improved sanitation, in percentage

- Less than 50%
- 50 to 75%
- 76 to 90%
- 91 to 100%
- No data

Source: World Health Organization

(*continued on page 54*)

influence on social relations, particularly in cultures that have retained strong connections to local environments. Changes in provisioning and regulating services can be mediated by socioeconomic factors, but those in cultural services cannot. Even a wealthy country like Sweden or the United Kingdom cannot readily purchase a substitute to a cultural landscape that is valued by the people in the community.

Changes in ecosystems have tended to increase the accessibility that people have to ecosystems for recreation and ecotourism. There are clear examples of declining ecosystem services disrupting social relations or resulting in conflicts. Indigenous societies whose cultural identities are tied closely to particular habitats or wildlife suffer if habitats are destroyed or wildlife populations decline. Such impacts have been observed in coastal fishing communities, Arctic populations, traditional forest societies, and pastoral nomadic societies (C5.4.4).

Security

By security, we refer to safety of person and possessions, secure access to necessary resources, and security from natural and human-made disasters. Changes in regulating services such as disease regulation, climate regulation, and flood regulation have very strong influences on security. Changes in provisioning services such as food and water have strong impacts on security, since degradation of these can lead to loss of access to these essential resources. Changes in cultural services can influence security since they can contribute to the breakdown or strengthening of social networks within society. Changes in supporting services have a strong influence by virtue of their influence on all the other categories of services. These benefits are moderately mediated by socioeconomic circumstances. The wealthy have access to some safety nets that can minimize the impacts of some ecosystem changes (such as flood or drought insurance). Nevertheless, the wealthy cannot entirely escape exposure to some of these changes in areas where they live.

One example of an aspect of security affected by ecosystem change involves influences on the severity and magnitude of floods

and major fires. The incidence of these has increased significantly over the past 50 years. Changes in ecosystems and in the management of ecosystems have contributed to these trends. The canalization of rivers, for example, tends to decrease the incidence and impact of small flood events and increase the incidence and severity of large ones. On average, 140 million people are affected by floods each year—more than all other natural or technological disasters put together. Between 1990 and 1999, more than 100,000 people were killed in floods, which caused a total of $243 billion in damages (C7.4.4).

Freedom of Choice and Action

Freedom of choice and action refers to the ability of individuals to control what happens to them and to be able to achieve what they value doing or being. Freedom and choice cannot exist without the presence of the other elements of well-being, so there is an indirect influence of changes in all categories of ecosystem services on the attainment of this constituent of well-being. The influence

of ecosystem change on freedom and choice is heavily mediated by socioeconomic circumstances. The wealthy and people living in countries with efficient governments and strong civil society can maintain freedom and choice even in the face of significant ecosystem change, while this would be impossible for the poor if, for example, the ecosystem change resulted in a loss of livelihood.

In the aggregate, the state of our knowledge about the impact that changing ecosystem conditions have on freedom and choice is relatively limited. Declining provision of fuelwood and drinking water have been shown to increase the amount of time needed to collect such basic necessities, which in turn reduces the amount of time available for education, employment, and care of family members. Such impacts are typically thought to be disproportionately experienced by women (although the empirical foundation for this view is relatively limited) (C5.4.2).

Figure 3.1. Net National Savings in 2001 Adjusted for Investments in Human Capital, Natural Resource Depletion, and Damage Caused by Pollution Compared with Standard Net National Savings Measurements (C5.2.6)

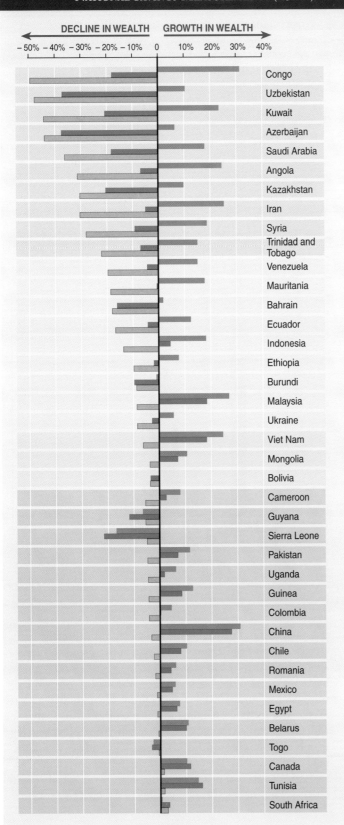

Positive values for national savings (expressed as a percent of gross national income) reflect a gain in wealth for a nation. Standard measures do not incorporate investments in human capital (in standard national accounting, these expenditures are treated as consumption), depletion of a variety of natural resources, or pollution damages. The World Bank provides estimates of adjusted net national savings, taking into account education expenses (which are added to standard measures), unsustainable forest harvest, depletion of nonrenewable resources (minerals and energy), and damage from carbon emissions related to its contribution to climate change (all of which are subtracted from the standard measure). The adjusted measure still overestimates actual net national savings, since it does not include potential changes in many ecosystem services including depletion of fisheries, atmospheric pollution, degradation of sources of fresh water, and loss of noncommercial forests and the ecosystem services they provide. Here we show the change in net national savings in 2001 for countries in which there was a decline of at least 5% in net national savings due to the incorporation of resource depletion or damage from carbon emissions.

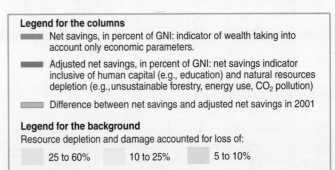

Source: Millennium Ecosystem Assessment

aesthetic benefits, do not pass through markets. The benefits they provide to society, therefore, are largely unrecorded: only a portion of the total benefits provided by an ecosystem make their way into statistics, and many of these are misattributed (the water regulation benefits of wetlands, for example, do not appear as benefits of wetlands but as higher profits in water-using sectors). Moreover, for ecosystem services that do not pass through markets there is often insufficient incentive for individuals to invest in maintenance (although in some cases common property management systems provide such incentives). Typically, even if individuals are aware of the services provided by an ecosystem, they are neither compensated for providing these services nor penalized for reducing them. These nonmarketed benefits are often high and sometimes more valuable than the marketed benefits. For example:

- *Total economic value of forests.* One of the most comprehensive studies to date, which examined the marketed and nonmarketed economic values associated with forests in eight Mediterranean countries, found that timber and fuelwood generally accounted for less than a third of total economic value in each country. (See Figure 3.2.)
- *Recreational benefits of protected areas:* The annual recreational value of the coral reefs of each of six Marine Management Areas in the Hawaiian Islands in 2003 ranged from $300,000 to $35 million.
- *Water quality:* The net present value in 1998 of protecting water quality in the 360-kilometer Catawba River in the United States for five years was estimated to be $346 million.
- *Water purification service of wetlands:* About half of the total economic value of the Danube River Floodplain in 1992 could be accounted for in its role as a nutrient sink.
- *Native pollinators:* A study in Costa Rica found that forest-based pollinators increased coffee yields by 20% within 1 kilometer of the forest (as well as increasing the quality of the coffee). During 2000–03, pollination services from two forest fragments (of 46 and 111 hectares) thus increased the income of a 1,100-hectare farm by $60,000 a year, a value commensurate with expected revenues from competing land uses.
- *Flood control:* Muthurajawela Marsh, a 3,100-hectare coastal peat bog in Sri Lanka, provides an estimated $5 million in annual benefits ($1,750 per hectare) through its role in local flood control.

■ *The total economic value associated with managing ecosystems more sustainably is often higher than the value associated with the conversion of the ecosystem through farming, clear-cut logging, or other intensive uses.* Relatively few studies have compared the total economic value (including values of both marketed and nonmarketed ecosystem services) of ecosystems under alternate management regimes, but a number of studies that do exist have found that the benefit of managing the ecosystem more sustainably exceeded that of converting the ecosystem (see Figure 3.3), although the private benefits—that is, the actual monetary benefits captured from the services entering the market—would favor conversion or unsustainable management. These studies are consistent with the understanding that market failures associated with ecosystem services lead to greater conversion of ecosystems than is economically justified. However, this finding would not hold at all locations. For example, the value of conversion of an ecosystem in areas of prime agricultural land or in urban regions often exceeds the total economic value of the intact ecosystem.

Figure 3.2. Annual Flow of Benefits from Forests in Selected Countries (Adapted from C5 Box 5.2)

In most countries, the marketed values of ecosystems associated with timber and fuelwood production are less than one third of the total economic value, including nonmarketed values such as carbon sequestration, watershed protection, and recreation.

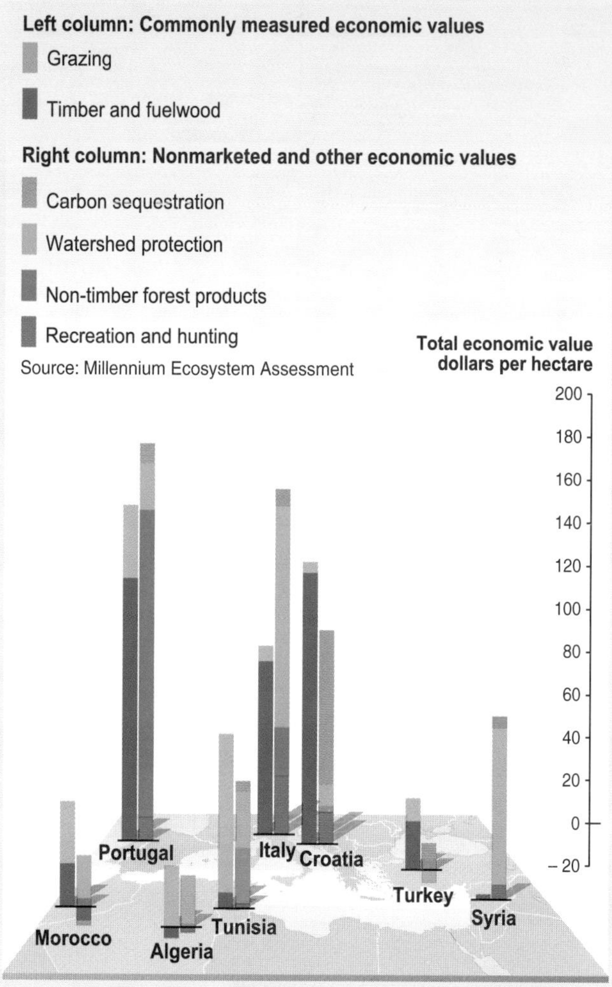
Left column: Commonly measured economic values
Grazing
Timber and fuelwood
Right column: Nonmarketed and other economic values
Carbon sequestration
Watershed protection
Non-timber forest products
Recreation and hunting
Source: Millennium Ecosystem Assessment
Total economic value dollars per hectare
200
180
160
140
120
100
80
60
40
20
0
– 20
Portugal
Italy
Croatia
Turkey
Syria
Tunisia
Morocco
Algeria

(Although even in dense urban areas, the total economic value of maintaining some "green space" can be greater than development of these sites.)

■ *The economic and public health costs associated with damage to ecosystem services can be substantial.*

- The early 1990s collapse of the Newfoundland cod fishery due to overfishing (see Figure 3.4) resulted in the loss of tens of thousands of jobs and has cost at least $2 billion in income support and retraining.
- The cost of U.K. agriculture in 1996 resulting from the damage that agricultural practices cause to water (pollution, eutrophication), air (emissions of greenhouse gases), soil (off-site erosion damage, carbon dioxide loss), and biodiversity was $2.6 billion, or 9% of average yearly gross farm receipts for the 1990s. Similarly, the damage costs of freshwater eutrophication alone in England and Wales was estimated to be $105–160 million per year in the 1990s, with an additional $77 million per year being spent to address those damages.
- The burning of 10 million hectares of Indonesia's forests in 1997/98 cost an estimated $9.3 billion in increased health care, lost production, and lost tourism revenues and affected some 20 million people across the region.
- The total damages for the Indian Ocean region over 20 years (with a 10% discount rate) resulting from the long-term impacts of the massive 1998 coral bleaching episode are estimated to be between $608 million (if there is only a slight decrease in tourism-generated income and employment results) and $8 billion (if tourism income and employment and fish productivity drop significantly and reefs cease to function as a protective barrier).
- The net annual loss of economic value associated with invasive species in the fynbos vegetation of the Cape Floral region of South Africa in 1997 was estimated to be $93.5 million, equivalent to a reduction of the potential economic value without the invasive species of more than 40%. The invasive species have caused losses of biodiversity, water, soil, and scenic beauty, although they also provide some benefits, such as provision of firewood.
- The incidence of diseases of marine organisms and emergence of new pathogens is increasing, and some of these, such as ciguatera, harm human health (C19.3.1). Episodes of harmful (including toxic) algal blooms in coastal

waters are increasing in frequency and intensity, harming other marine resources such as fisheries and harming human health (R16 Figure 16.3). In a particularly severe outbreak in Italy in 1989, harmful algal blooms cost the coastal aquaculture industry $10 million and the Italian tourism industry $11.4 million (C19.3.1).

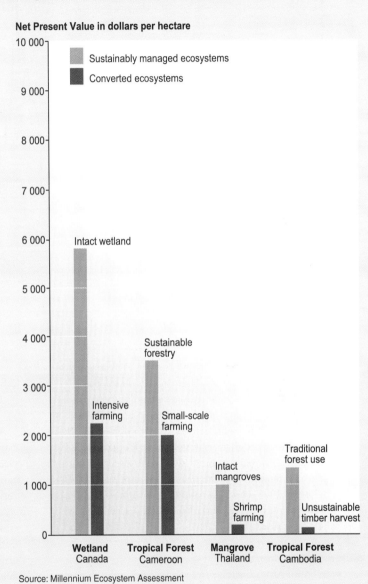

Figure 3.3. Economic Benefits under Alternate Management Practices (C5 Box 5.2)

In each case, the net benefits from the more sustainably managed ecosystem are greater than those from the converted ecosystem even though the private (market) benefits would be greater from the converted ecosystem. (Where ranges of values are given in the original source, lower estimates are plotted here.)

Net Present Value in dollars per hectare

Source: Millennium Ecosystem Assessment

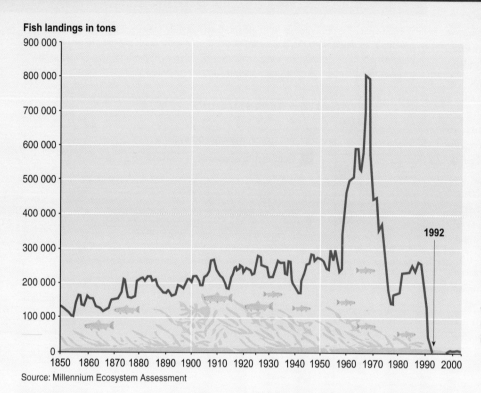

Fish landings in tons

Source: Millennium Ecosystem Assessment

This collapse forced the closure of the fishery after hundreds of years of exploitation. Until the late 1950s, the fishery was exploited by migratory seasonal fleets and resident inshore small-scale fishers. From the late 1950s, offshore bottom trawlers began exploiting the deeper part of the stock, leading to a large catch increase and a strong decline in the underlying biomass. Internationally agreed quotas in the early 1970s and, following the declaration by Canada of an Exclusive Fishing Zone in 1977, national quota systems ultimately failed to arrest and reverse the decline. The stock collapsed to extremely low levels in the late 1980s and early 1990s, and a moratorium on commercial fishing was declared in June 1992. A small commercial inshore fishery was reintroduced in 1998, but catch rates declined and the fishery was closed indefinitely in 2003.

■ The number of both floods and fires has increased significantly, in part due to ecosystem changes, in the past 50 years. Examples are the increased susceptibility of coastal populations to tropical storms when mangrove forests are cleared and the increase in downstream flooding that followed land use changes in the upper Yangtze river (C.SDM). Annual economic losses from extreme events increased tenfold from the 1950s to approximately $70 billion in 2003, of which natural catastrophes—floods, fires, storms, drought, and earthquakes—accounted for 84% of insured losses.

■ *Significant investments are often needed to restore or maintain nonmarketed ecosystem services.*

■ In South Africa, invasive tree species threaten both native species and water flows by encroaching into natural habitats, with serious impacts for economic growth and human well-being. In response, the South African government established the "Working for Water Programme." Between 1995 and 2001 the program invested $131 million (at 2001 exchange rates) in clearing programs to control the invasive species.

■ The state of Louisiana has put in place a $14-billion wetland restoration plan to protect 10,000 square kilometers of marsh, swamp, and barrier islands in part to reduce storm surges generated by hurricanes.

Although degradation of ecosystem services could be significantly slowed or reversed if the full economic value of the services were taken into account in decision-making, economic considerations alone would likely lead to lower levels of biodiversity *(medium certainty)* **(CWG).** Although most or all biodiversity has some economic value (the option value of any species is always greater than zero), that does not mean that the protection of all biodiversity is always economically justified. Other utilitarian benefits often "compete" with the benefits of maintaining greater diversity. For example, many of the steps taken to increase the production of ecosystem services involve the simplification of natural systems. (Agriculture, for instance, typically has involved the replacement of relatively diverse systems with more simplified production systems.) And protecting some other ecosystem services may not necessarily require the conservation of biodiversity. (For example, a forested watershed could provide clean water whether it was covered in a diverse native forest or in a single-species plantation.) Ultimately, the level of biodiversity that survives on Earth will be determined not just by utilitarian considerations but to a significant extent by ethical concerns, including considerations of the intrinsic values of species.

Even wealthy populations cannot be fully insulated from the degradation of ecosystem services (CWG). The degradation of ecosystem services influences human well-being in industrial regions as well as wealthy populations in developing countries.

- The physical, economic, or social impacts of ecosystem service degradation may cross boundaries. (See Figure 3.5.) Land degradation and fires in poor countries, for example, have contributed to air quality degradation (dust and smoke) in wealthy ones.
- Degradation of ecosystem services exacerbates poverty in developing countries, which can affect neighboring industrial countries by slowing regional economic growth and contributing to the outbreak of conflicts or the migration of refugees.
- Changes in ecosystems that contribute to greenhouse gas emissions contribute to global climate changes that affect all countries.
- Many industries still depend directly on ecosystem services. The collapse of fisheries, for example, has harmed many communities in industrial countries. Prospects for the forest, agriculture, fishing, and ecotourism industries are all directly tied to ecosystem services, while other sectors such as insurance, banking, and health are strongly, if less directly, influenced by changes in ecosystem services.
- Wealthy populations are insulated from the harmful effects of some aspects of ecosystem degradation, but not all. For example, substitutes are typically not available when cultural services are lost.

While traditional natural resource sectors such as agriculture, forestry, and fisheries are still important in industrial-country economies, the relative economic and political significance of other sectors has grown as a result of the ongoing transition from agricultural to industrial and service economies (S7). Over the past two centuries, the economic structure of the world's largest economies has shifted significantly from agricultural production to industry and, in particular, to service industries. (See Figure 3.6.) These changes increase the relative significance of the industrial and service sectors (using conventional economic measures that do not factor in nonmarketed costs and benefits) in comparison to agriculture, forestry, and fisheries, although natural resource–based sectors often still dominate in developing countries. In 2000, agriculture accounted for 5% of gross world product, industry 31%, and service industries 64%. At the same time, the importance of other nonmarketed ecosystem services has grown, although many of the benefits provided by these services are not captured in national economic statistics. The economic value of water from forested ecosystems near urban populations, for example, now sometimes exceeds the value of timber in those ecosystems. Economic and employment contributions from ecotourism, recreational hunting, and fishing have all grown.

Increased trade has often helped meet growing demand for ecosystem services such as grains, fish, and timber in regions where their supply is limited. While this lessens pressures on

Figure 3.5. Dust Cloud off the Northwest Coast of Africa, March 6, 2004

In this image, the storm covers about one fifth of Earth's circumference. The dust clouds travel thousands of miles and fertilize the water off the west coast of Florida with iron. This has been linked to blooms of toxic algae in the region and respiratory problems in North America and has affected coral reefs in the Caribbean. Degradation of drylands exacerbates problems associated with dust storms.

Source: National Aeronautics and Space Administration, Earth Observatory

ecosystem services within the importing region, it increases pressures in the exporting region. Fish products are heavily traded, and approximately 50% of exports are from developing countries. Exports from these nations and the Southern Hemisphere presently offset much of the shortfall of supply in European, North American, and East Asian markets (C18.ES). Trade has increased the quantity and quality of fish supplied to wealthy countries, in particular the United States, those in Europe, and Japan, despite reductions in marine fish catch (C18.4.1).

The value of international trade in forest products has increased much faster than increases in harvests. (Roundwood harvests grew by 60% between 1961 and 2000, while the value of international timber trade increased twenty-five-fold (C9.ES).) The United States, Germany, Japan, United Kingdom, and Italy were the destination of more than half of the imports in 2000, while Canada, United States, Sweden, Finland, and Germany account for more than half of the exports.

Trade in commodities such as grain, fish, and timber is accompanied by a "virtual trade" in other ecosystem services that are required to support the production of these commodities.

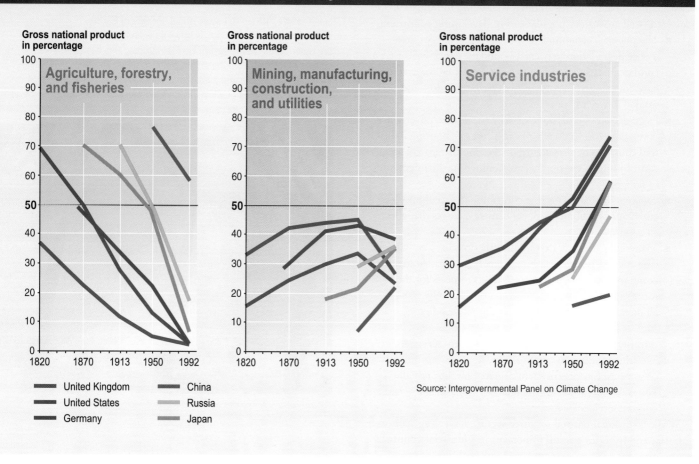

Figure 3.6. CHANGES IN ECONOMIC STRUCTURE FOR SELECTED COUNTRIES. This indicates the share of national GDP for different sectors between 1820 and 1992. (S7 Fig 7.3)

Gross national product in percentage

Agriculture, forestry, and fisheries

Gross national product in percentage

Mining, manufacturing, construction, and utilities

Gross national product in percentage

Service industries

Source: Intergovernmental Panel on Climate Change

United Kingdom
United States
Germany
China
Russia
Japan

Globally, the international virtual water trade in crops has been estimated between 500 and 900 cubic kilometers per year, and 130–150 cubic kilometers per year is traded in livestock and livestock products. For comparison, current rates of water consumption for irrigation total 1,200 cubic kilometers per year (C7.3.2).

Changes in ecosystem services affect people living in urban ecosystems both directly and indirectly. Likewise, urban populations have strong impacts on ecosystem services both in the local vicinity and at considerable distances from urban centers (C27). Almost half of the world's population now lives in urban areas, and this proportion is growing. Urban development often threatens the availability of water, air and water quality, waste processing, and many other qualities of the ambient environment that contribute to human well-being, and this degradation is particularly threatening to vulnerable groups such as poor people. A wide range of ecosystem services are still important to livelihoods. For example, agriculture practiced within urban boundaries contributes to food security in urban sub-Saharan Africa.

Urban populations affect distant ecosystems through trade and consumption and are affected by changes in distant ecosystems that affect the local availability or price of commodities, air or water quality, or global climate, or that affect socioeconomic conditions in those countries in ways that influence the economy, demographic, or security situation in distant urban areas.

Spiritual and cultural values of ecosystems are as important as other services for many local communities. Human cultures, knowledge systems, religions, heritage values, and social interactions have always been influenced and shaped by the nature of the ecosystem and ecosystem conditions in which culture is based. People have benefited in many ways from cultural ecosystem services, including aesthetic enjoyment, recreation, artistic and spiritual fulfillment, and intellectual development (C17.ES). Several of the MA sub-global assessments highlighted the importance of these cultural services and spiritual benefits to local communities (SG.SDM). For example, local villages in India preserve selected sacred groves of forest for spiritual reasons, and urban parks provide important cultural and recreational services in cities around the world.

Ecosystem Services, Millennium Development Goals, and Poverty Reduction

The degradation of ecosystem services poses a significant barrier to the achievement of the Millennium Development Goals and to the MDG targets for 2015. (See Box 3.2.) Many of the regions facing the greatest challenges in achieving the MDGs overlap with the regions facing the greatest problems related to the sustainable supply of ecosystem services (R19.ES). Among other regions, this includes sub-Saharan Africa, Central Asia, and parts of South and Southeast Asia as well as some regions in Latin America. Sub-Saharan Africa has experienced increases in maternal deaths and income poverty (those living on less than $1 a day), and the number of people living in poverty there is forecast to rise from 315 million in 1999 to 404 million by 2015 (R19.1). Per capita food production has been declining in southern Africa, and relatively little gain is projected in the MA scenarios. Many of these regions include large areas of drylands, in which a combination of growing populations and land degradation are increasing the vulnerability of people to both economic and environmental change. In the past 20 years, these same regions have experienced some of the highest rates of forest and land degradation in the world.

Despite the progress achieved in increasing the production and use of some ecosystem services, levels of poverty remain high, inequities are growing, and many people still do not have a sufficient supply of or access to ecosystem services (C5).

- In 2001, some 1.1 billion people survived on less than $1 per day of income, most of them (roughly 70%) in rural areas where they are highly dependent on agriculture, grazing, and hunting for subsistence (R19.2.1).
- Inequality in income and other measures of human well-being has increased over the past decade (C5.ES). A child born in sub-Saharan Africa is 20 times more likely to die before age five than a child born in an industrial country, and this ratio is higher than it was a decade ago. During the 1980s, only four countries experienced declines in their rankings in the Human Development Index (an aggregate measure of economic well-being, health, and education); during the 1990s, 21 countries showed declines, and 14 of them were in sub-Saharan Africa.
- Despite the growth in per capita food production in the past four decades, an estimated 852 million people were undernourished in 2000–02, up 37 million from 1997–99. Of these, nearly 95% live in developing countries (C8.ES).

BOX 3.2. ECOSYSTEMS AND THE MILLENNIUM DEVELOPMENT GOALS

The eight Millennium Development Goals were endorsed by governments at the United Nations in September 2000. The MDGs aim to improve human well-being by reducing poverty, hunger, and child and maternal mortality; ensuring education for all; controlling and managing diseases; tackling gender disparity; ensuring sustainable development; and pursuing global partnerships. For each MDG, governments have agreed to between 1 and 8 targets (a total of 15 targets) that are to be achieved by 2015. Slowing or reversing the degradation of ecosystem services will contribute significantly to the achievement of many of the MDGs.

■ **Poverty Eradication.** Ecosystem services are a dominant influence on livelihoods of most poor people. Most of the world's poorest people live in rural areas and are thus highly dependent, directly or indirectly, on the ecosystem service of food production, including agriculture, livestock, and hunting (R19.2.1). Mismanagement of ecosystems threatens the livelihood of poor people and may threaten their survival (C5.ES). Poor people are highly vulnerable to changes in watershed services that affect the quality or availability of water, loss of ecosystems such as wetlands, mangroves, or coral

reefs that affect the likelihood of flood or storm damage, or changes in climate regulating services that might alter regional climate. Ecosystem degradation is often one of the factors trapping people in cycles of poverty.

■ **Hunger Eradication** (R19.2.2). Although economic and social factors are often the primary determinants of hunger, food production remains an important factor, particularly among the rural poor. Food production is an ecosystem service in its own right, and it also depends on watershed services, pollination, pest regulation, and soil formation. Food production needs to increase to meet the needs of the growing human population, and at the same time the efficiency of food production (the amount produced per unit of land, water, and other inputs) needs to increase in order to reduce harm to other key ecosystem services. Ecosystem condition, in particular climate, soil degradation, and water availability, influences progress toward this goal through its influence on crop yields as well as through impacts on the availability of wild sources of food.

■ **Reducing Child Mortality.** Undernutrition is the underlying cause of a substantial proportion of all child deaths. Child mortality is also strongly influenced by diseases associ-

ated with water quality. Diarrhea is one of the predominant causes of infant deaths worldwide. In sub-Saharan Africa, malaria additionally plays an important part in child mortality in many countries of the region.

■ **Combating Disease** (R19.2.7). Human health is strongly influenced by ecosystem services related to food production, water quality, water quantity, and natural hazard regulation, and the role of ecosystem management is central to addressing some of the most pressing global diseases such as malaria. Changes in ecosystems influence the abundance of human pathogens such as malaria and cholera as well as the risk of emergence of new diseases. Malaria is responsible for 11% of the disease burden in Africa, and it is estimated that Africa's GDP could have been $100 billion larger (roughly a 25% increase) in 2000 if malaria had been eliminated 35 years ago (R16.1).

■ **Environmental Sustainability.** Achievement of this goal will require, at a minimum, an end to the current unsustainable uses of ecosystem services such as fisheries and fresh water and an end to the degradation of other services such as water purification, natural hazard regulation, disease regulation, climate regulation, and cultural amenities.

South Asia and sub-Saharan Africa, the regions with the largest numbers of undernourished people, are also the regions where growth in per capita food production has lagged the most. Most notably, per capita food production has declined in sub-Saharan Africa (C28.5.1).

- Some 1.1 billion people still lack access to improved water supply and more than 2.6 billion have no access to improved sanitation. Water scarcity affects roughly 1–2 billion people worldwide. Since 1960, the ratio of water use to accessible supply has grown by 20% per decade (C7.ES, C7.2.3).

The degradation of ecosystem services is harming many of the world's poorest people and is sometimes the principal factor causing poverty. This is not to say that ecosystem changes such as increased food production have not also helped to lift hundreds of millions of people out of poverty. But these changes have harmed many other communities, and their plight has been largely overlooked. Examples of these impacts include:

- Half of the urban population in Africa, Asia, Latin America, and the Caribbean suffers from one or more diseases associated with inadequate water and sanitation (C.SDM). Approximately 1.7 million people die annually as a result of inadequate water, sanitation, and hygiene (C7.ES).
- The declining state of capture fisheries is reducing a cheap source of protein in developing countries. Per capita fish consumption in developing countries, excluding China, declined between 1985 and 1997 (C18.ES).
- Desertification affects the livelihoods of millions of people, including a large portion of the poor in drylands (C22).

The pattern of "winners" and "losers" associated with ecosystem changes, and in particular the impact of ecosystem changes on poor people, women, and indigenous peoples, has not been adequately taken into account in management decisions (R17). Changes in ecosystems typically yield benefits for some people and exact costs on others, who may either lose access to resources or livelihoods or be affected by externalities associated with the change. For several reasons, groups such as the poor, women, and indigenous communities have tended to be harmed by these changes.

- Many changes have been associated with the privatization of what were formerly common pool resources, and the individuals who are dependent on those resources have thus lost rights to them. This has been particularly the case for indigenous peoples, forest-dependent communities, and other groups relatively marginalized from political and economic sources of power.
- Some of the people and places affected by changes in ecosystems and ecosystem services are highly vulnerable and poorly equipped to cope with the major ecosystem changes that may occur (C6.ES). Highly vulnerable groups include

those whose needs for ecosystem services already exceed the supply, such as people lacking adequate clean water supplies and people living in areas with declining per capita agricultural production. Vulnerability has also been increased by the growth of populations in ecosystems at risk of disasters such as floods or drought, often due to inappropriate policies that have encouraged this growth. Populations are growing in low-lying coastal areas and dryland ecosystems. In part due to the growth in these vulnerable populations, the number of natural disasters (floods, droughts, earthquakes, and so on) requiring international assistance has quadrupled over the past four decades. Finally, vulnerability has been increased when the resilience in either the social or ecological system has been diminished, as for example through the loss of drought-resistant crop varieties.

- Significant differences between the roles and rights of men and women in many societies lead to women's increased vulnerability to changes in ecosystem services. Rural women in developing countries are the main producers of staple crops like rice, wheat, and maize (R6 Box 6.1). Because the gendered division of labor within many societies places responsibility for routine care of the household with women, even when women also play important roles in agriculture, the degradation of ecosystem services such as water quality or quantity, fuelwood, agricultural or rangeland productivity often results in increased labor demands on women. This can affect the larger household by diverting time from food preparation, child care, education of children, and other beneficial activities (C6.3.3). Yet gender bias persists in agricultural policies in many countries, and rural women involved in agriculture tend to be the last to benefit from—or in some cases are negatively affected by—development policies and new technologies.
- The reliance of the rural poor on ecosystem services is rarely measured and thus typically overlooked in national statistics and in poverty assessments, resulting in inappropriate strategies that do not take into account the role of the environment in poverty reduction. For example, a recent study that synthesized data from 17 countries found that 22% of household income for rural communities in forested regions comes from sources typically not included in national statistics, such as harvesting wild food, fuelwood, fodder, medicinal plants, and timber. These activities generated a much higher proportion of poorer families' total income than wealthy families'—income that was of particular significance in periods of both predictable and unpredictable shortfalls in other livelihood sources (R17).

Poor people have historically lost access to ecosystem services disproportionately as demand for those services has grown. Coastal habitats are often converted to other uses, frequently for aquaculture ponds or cage culturing of highly valued species such as shrimp and salmon. Despite the fact that the area is still used for food production, local residents are often displaced, and the

food produced is usually not for local consumption but for export (C18.4.1). Many areas where overfishing is a concern are also low-income, food-deficit countries. For example, significant quantities of fish are caught by large distant water fleets in the exclusive economic zones of Mauritania, Senegal, Gambia, Guinea Bissau, and Sierra Leone. Much of the catch is exported or shipped directly to Europe, while compensation for access is often low compared with the value of the product landed overseas. These countries do not necessarily benefit through increased fish supplies or higher government revenues when foreign distant water fleets ply their waters (C18.5.1).

Diminished human well-being tends to increase immediate dependence on ecosystem services, and the resultant additional pressure can damage the capacity of those ecosystems to deliver services (SG3.ES). As human well-being declines, the options available to people that allow them to regulate their use of natural resources at sustainable levels decline as well. This in turn increases pressure on ecosystem services and can create a downward spiral of increasing poverty and further degradation of ecosystem services.

Dryland ecosystems tend to have the lowest levels of human well-being (C5.3.3). Drylands have the lowest per capita GDP and the highest infant mortality rates of all of the MA systems Nearly 500 million people live in rural areas in dry and semiarid lands, mostly in Asia and Africa but also in regions of Mexico and northern Brazil (C5 Box 5.2). The small amount of precipitation and its high variability limit the productive potential of drylands for settled farming and nomadic pastoralism, and many

ways of expanding production (such as reducing fallow periods, overgrazing pasture areas, and cutting trees for fuelwood) result in environmental degradation. The combination of high variability in environmental conditions and relatively high levels of poverty leads to situations where human populations can be extremely sensitive to changes in the ecosystem (although the presence of these conditions has led to the development of very resilient land management strategies). Once rainfall in the Sahel reverted to normal low levels after 1970, following favorable rainfall from the 1950s to the mid-1960s that had attracted people to the region, an estimated 250,000 people died, along with nearly all their cattle, sheep, and goats (C5 Box 5.1).

Although population growth has historically been higher in high-productivity ecosystems or urban areas, during the 1990s it was highest in less productive ecosystems (C5.ES, C5.3.4). In that decade dryland systems (encompassing both rural and urban regions of drylands) experienced the highest, and mountain systems the second highest, population growth rate of any of the systems examined in the MA. (See Figure 3.7.) One factor that has helped reduce relative population growth in marginal lands has been migration of some people out of marginal lands to cities or to agriculturally productive regions; today the opportunities for such migration are limited due to a combination of factors, including poor economic growth in some cities, tighter immigration restrictions in wealthy countries, and limited availability of land in more productive regions.

Figure 3.7. HUMAN POPULATION GROWTH RATES, 1990–2000, AND PER CAPITA GDP AND BIOLOGICAL PRODUCTIVITY IN 2000 IN MA ECOLOGICAL SYSTEMS

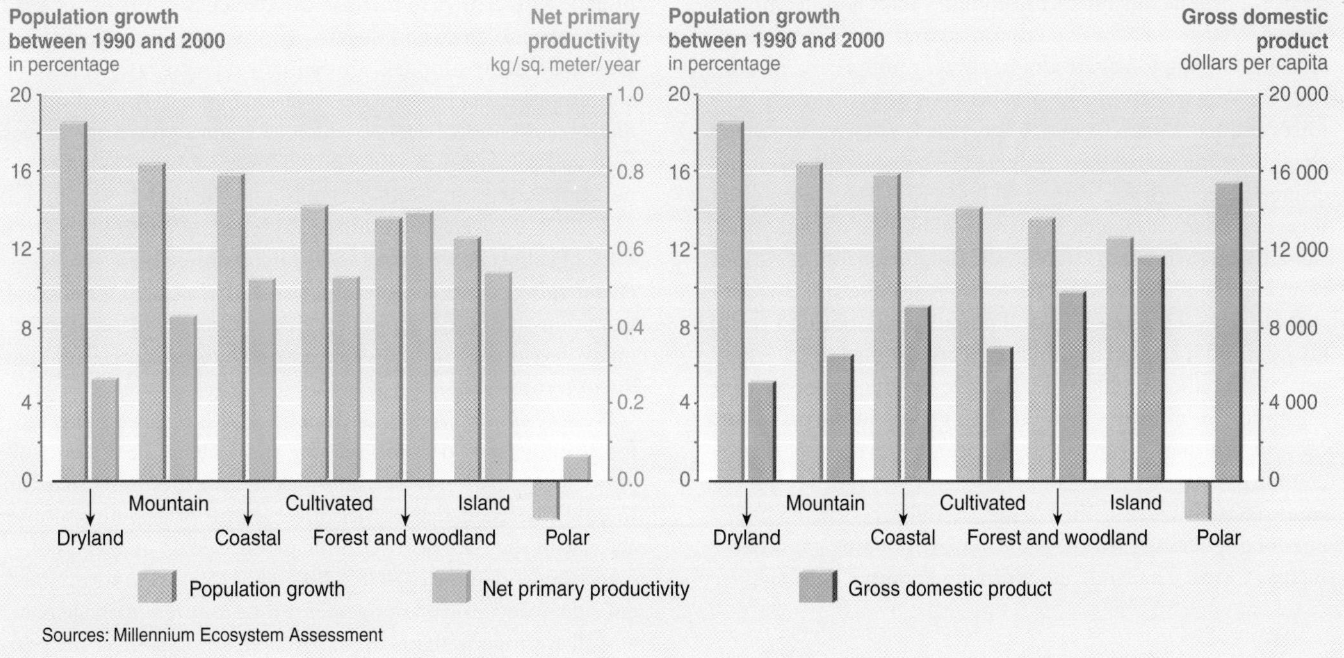

Sources: Millennium Ecosystem Assessment

4. *What are the most critical factors causing ecosystem changes?*

Natural or human-induced factors that directly or indirectly cause a change in an ecosystem are referred to as "drivers." A *direct driver* unequivocally influences ecosystem processes. An *indirect driver* operates more diffusely, by altering one or more direct drivers.

Drivers affect ecosystem services and human well-being at different spatial and temporal scales, which makes both their assessment and their management complex (SG7). Climate change may operate on a global or a large regional spatial scale; political change may operate at the scale of a nation or a municipal district. Sociocultural change typically occurs slowly, on a time scale of decades (although abrupt changes can sometimes occur, as in the case of wars or political regime changes), while economic changes tend to occur more rapidly. As a result of this spatial and temporal dependence of drivers, the forces that appear to be most significant at a particular location and time may not be the most significant over larger (or smaller) regions or time scales.

Indirect Drivers

In the aggregate and at a global scale, there are five indirect drivers of changes in ecosystems and their services: population change, change in economic activity, sociopolitical factors, cultural factors, and technological change. Collectively these factors influence the level of production and consumption of ecosystem services and the sustainability of production. Both economic growth and population growth lead to increased consumption of ecosystem services, although the harmful environmental impacts of any particular level of consumption depend on the efficiency of the technologies used in the production of the service. These factors interact in complex ways in different locations to change pressures on ecosystems and uses of ecosystem services. Driving forces are almost always multiple and interactive, so that a one-to-one linkage between particular driving forces and particular changes in ecosystems rarely exists. Even so, changes in any one of these indirect drivers generally result in changes in ecosystems. The causal linkage is almost always highly mediated by other factors, thereby complicating statements of causality or attempts to establish the proportionality of various contributors to changes. There are five major indirect drivers:

■ *Demographic Drivers:* Global population doubled in the past 40 years and increased by 2 billion people in the last 25 years, reaching 6 billion in 2000 (S7.2.1). Developing countries have accounted for most recent population growth in the past quarter-century, but there is now an unprecedented diversity of demographic patterns across regions and countries. Some high-income countries such as the United States are still experiencing high rates of population growth, while some developing countries such as China, Thailand, and North and South Korea have very low rates. In the United States, high population growth is due primarily to high levels of immigration. About half the people in the world now live in urban areas (although urban areas cover less than 3% of the terrestrial surface), up from less than 15% at the start of the twentieth century (C27.1). High-income countries typically have populations that are 70–80% urban. Some developing-country regions, such as parts of Asia, are still largely rural, while Latin America, at 75% urban, is indistinguishable from high-income countries in this regard (S7.2.1).

■ *Economic Drivers:* Global economic activity increased nearly sevenfold between 1950 and 2000 (S7.SDM). With rising per capita income, the demand for many ecosystem services grows. At the same time, the structure of consumption changes. In the case of food, for example, as income grows the share of additional income spent on food declines, the importance of starchy staples (such as rice, wheat, and potatoes) declines, diets include more fat, meat and fish, and fruits and vegetables, and the proportionate consumption of industrial goods and services rises (S7.2.2).

In the late twentieth century, income was distributed unevenly, both within countries and around the world. The level of per capita income was highest in North America, Western Europe, Australasia, and Northeast Asia, but both GDP growth rates and per capita GDP growth rates were highest in South Asia, China, and parts of South America (S7.2.2). (See Figures 4.1 and 4.2.) Growth in international trade flows has exceeded growth in global production for many years, and the differential may be growing. In 2001, international trade in goods was equal to 40% of gross world product. (S7.2.2).

Taxes and subsidies are important indirect drivers of ecosystem change. Fertilizer taxes or taxes on excess nutrients, for example, provide an incentive to increase the efficiency of the use of fertilizer applied to crops and thereby reduce negative externalities. Currently, many subsidies substantially increase rates of resource consumption and increase negative externalities. Annual subsidies to conventional energy, which encourage greater use of fossil fuels and consequently emissions of greenhouse gases, are estimated to have been $250–300 billion in the mid-1990s (S7.ES). The 2001–03 average subsidies paid to the agricultural sectors of OECD countries were over $324 billion annually (S7.ES), encouraging greater food production and associated water consumption and nutrient and pesticide release. At the same time, many developing countries also have significant agricultural production subsidies.

■ *Sociopolitical Drivers:* Sociopolitical drivers encompass the forces influencing decision-making and include the quantity of public participation in decision-making, the groups participating in public decision-making, the mechanisms of dispute resolution, the role of the state relative to the private sector, and levels of education and knowledge (S7.2.3). These factors in turn influence the institutional arrangements for ecosystem management, as well as property rights over ecosystem services. Over the past

Figure 4.1. GDP Average Annual Growth, 1990–2003 (S7 Fig 7.6b)

Average annual percentage growth rate of GDP at market prices based on constant local currency. Dollar figures for GDP are converted from domestic currencies using 1995 official exchange rates. GDP is the sum of gross value added by all resident producers in the economy plus any product taxes and minus any subsidies not included in the value of the products. It is calculated without making deductions for depreciation of fabricated assets or for depletion and degradation of natural resources.

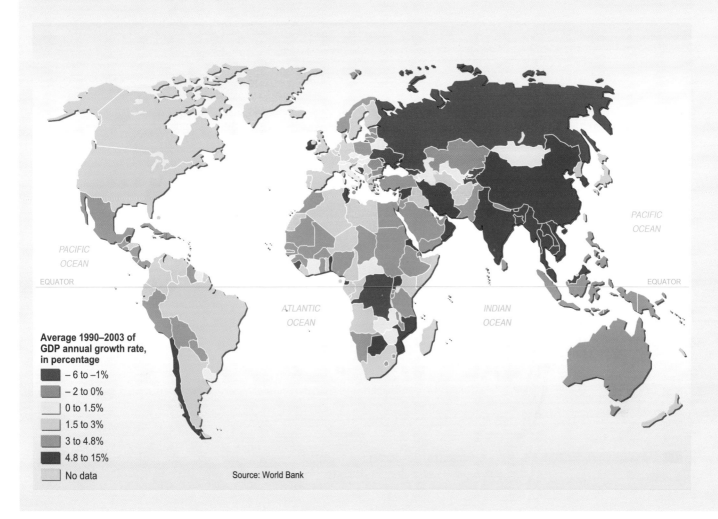

Average 1990–2003 of GDP annual growth rate, in percentage

- – 6 to –1%
- – 2 to 0%
- 0 to 1.5%
- 1.5 to 3%
- 3 to 4.8%
- 4.8 to 15%
- No data

Source: World Bank

50 years there have been significant changes in sociopolitical drivers. There is a declining trend in centralized authoritarian governments and a rise in elected democracies. The role of women is changing in many countries, average levels of formal education are increasing, and there has been a rise in civil society (such as increased involvement of NGOs and grassroots organizations in decision-making processes). The trend toward democratic institutions has helped give power to local communities, especially women and resource-poor households (S7.2.3). There has been an increase in multilateral environmental agreements. The importance of the state relative to the private sector—as a supplier of goods and services, as a source of employment, and as a source of innovation—is declining.

■ *Cultural and Religious Drivers:* To understand culture as a driver of ecosystem change, it is most useful to think of it as the values, beliefs, and norms that a group of people share. In this sense, culture conditions individuals' perceptions of the world, influences what they consider important, and suggests what courses of action are appropriate and inappropriate (S7.2.4). Broad comparisons of whole cultures have not proved useful because they ignore vast variations in values, beliefs, and norms within cultures. Nevertheless, cultural differences clearly have important impacts on direct drivers. Cultural factors, for example, can influence consumption behavior (what and how much people consume) and values related to environmental stewardship, and they may be particularly important drivers of environmental change.

Figure 4.2. PER CAPITA GDP AVERAGE ANNUAL GROWTH, 1990–2003 (S7 Fig 7.6a)

Average annual percentage growth rate of GDP per capita at market prices based on constant local currency. Dollar figures for GDP are converted from domestic currencies using 1995 official exchange rates. GDP is the sum of gross value added by all resident producers in the economy plus any product taxes and minus any subsidies not included in the value of the products. It is calculated without making deductions for depreciation of fabricated assets or for depletion and degradation of natural resources.

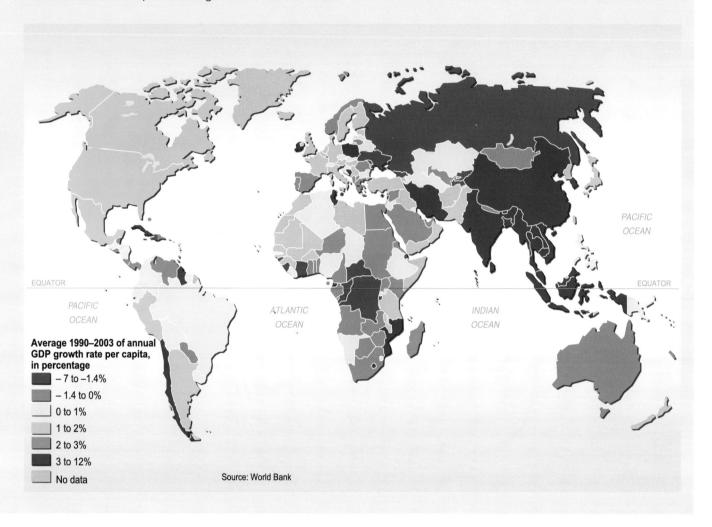

Average 1990–2003 of annual GDP growth rate per capita, in percentage

- − 7 to −1.4%
- − 1.4 to 0%
- 0 to 1%
- 1 to 2%
- 2 to 3%
- 3 to 12%
- No data

Source: World Bank

■ *Science and Technology:* The development and diffusion of scientific knowledge and technologies that exploit that knowledge has profound implications for ecological systems and human well-being. The twentieth century saw tremendous advances in understanding how the world works physically, chemically, biologically, and socially and in the applications of that knowledge to human endeavors. Science and technology are estimated to have accounted for more than one third of total GDP growth in the United States from 1929 to the early 1980s, and for 16–47% of GDP growth in selected OECD countries in 1960–95 (S7.2.5). The impact of science and technology on ecosystem services is most evident in the case of food production. Much of the increase in agricultural output over the past 40 years has come from an increase in yields per hectare rather than an expansion of area under cultivation. For instance, wheat yields rose 208%, rice yields rose 109%, and maize yields rose 157% in the past 40 years in developing countries (S7.2.5). At the same time, technological advances can also lead to the degradation of ecosystem services. Advances in fishing technologies, for example, have contributed significantly to the depletion of marine fish stocks.

Consumption of ecosystem services is slowly being decoupled from economic growth. Growth in the use of ecosystem services over the past five decades was generally much less than the growth in GDP. This change reflects structural changes in economies, but it also results from new technologies and new management practices and policies that have increased the efficiency with which ecosystem services are used and provided

substitutes for some services. Even with this progress, though, the absolute level of consumption of ecosystem services continues to grow, which is consistent with the pattern for the consumption of energy and materials such as metals: in the 200 years for which reliable data are available, growth of consumption of energy and materials has outpaced increases in materials and energy efficiency, leading to absolute increases of materials and energy use (S7.ES).

Global trade magnifies the effect of governance, regulations, and management practices on ecosystems and their services, enhancing good practices but worsening the damage caused by poor practices (R8, S7). Increased trade can accelerate degradation of ecosystem services in exporting countries if their policy, regulatory, and management systems are inadequate. At the same time, international trade enables comparative advantages to be exploited and accelerates the diffusion of more-efficient technologies and practices. For example, the increased demand for forest products in many countries stimulated by growth in forest products trade can lead to more rapid degradation of forests in countries with poor systems of regulation and management, but can also stimulate a "virtuous cycle" if the regulatory framework is sufficiently robust to prevent resource degradation while trade, and profits, increase. While historically most trade related to ecosystems has involved provisioning services such as food, timber, fiber, genetic resources, and biochemicals, one regulating service—climate regulation, or more specifically carbon sequestration—is now also traded internationally.

Urban demographic and economic growth has been increasing pressures on ecosystems globally, but affluent rural and suburban living often places even more pressure on ecosystems (C27.ES). Dense urban settlement is considered to be less environmentally burdensome than urban and suburban sprawl. And the movement of people into urban areas has significantly lessened pressure on some ecosystems and, for example, has led to the reforestation of some parts of industrial countries that had been deforested in previous centuries. At the same time, urban centers facilitate human access to and management of ecosystem services through, for example, economies of scale related to the construction of piped water systems in areas of high population density.

Direct Drivers

Most of the direct drivers of change in ecosystems and biodiversity currently remain constant or are growing in intensity in most ecosystems. (See Figure 4.3.) The most important direct drivers of change in ecosystems are habitat change (land use change and physical modification of rivers or water withdrawal from rivers), overexploitation, invasive alien species, pollution, and climate change.

For terrestrial ecosystems, the most important direct drivers of change in ecosystem services in the past 50 years, in the aggregate, have been land cover change (in particular, conversion to cropland) and the application of new technologies (which have contributed significantly to the increased supply of services such as food, timber, and fiber) (CWG, S7.2.5, SG8.ES). In 9 of the 14 terrestrial biomes examined in the MA, between one half and one fifth of the area has been transformed, largely to croplands (C4.ES). Only biomes relatively unsuited to crop plants, such as deserts, boreal forests, and tundra, have remained largely untransformed by human action. Both land cover changes and the management practices and technologies used on lands may cause major changes in ecosystem services. New technologies have resulted in significant increases in the supply of some ecosystem services, such as through increases in agricultural yield. In the case of cereals, for example, from the mid-1980s to the late 1990s the global area under cereals fell by around 0.3% a year, while yields increased by about 1.2% a year (C26.4.1).

For marine ecosystems and their services, the most important direct driver of change in the past 50 years, in the aggregate, has been fishing (C18). At the beginning of the twenty-first century, the biological capability of commercially exploited fish stocks was probably at a historical low. FAO estimates that about half of the commercially exploited wild marine fish stocks for which information is available are fully exploited and offer no scope for increased catches, and a further quarter are over exploited. (C8.2.2). As noted in Key Question 1, fishing pressure is so strong in some marine systems that the biomass of some targeted species, especially larger fishes, and those caught incidentally has been reduced to one tenth of levels prior to the onset of industrial fishing (C18.ES). Fishing has had a particularly significant impact in coastal areas but is now also affecting the open oceans.

For freshwater ecosystems and their services, depending on the region, the most important direct drivers of change in the past 50 years include modification of water regimes, invasive species, and pollution, particularly high levels of nutrient loading. It is *speculated* that 50% of inland water ecosystems (excluding large lakes and closed seas) were converted during the twentieth century (C20.ES). Massive changes have been made in water regimes: in Asia, 78% of the total reservoir volume was constructed in the last decade, and in South America almost 60% of all reservoirs have been built since the 1980s (C20.4.2). The introduction of non-native invasive species is one of the major causes of species extinction in freshwater systems. While the presence of nutrients such as phosphorus and nitrogen is necessary for biological systems, high levels of nutrient loading cause significant eutrophication of water bodies and contribute to high levels of nitrate in drinking water in some locations. (The nutrient load refers to the total amount of nitrogen or phosphorus entering the water during a given time.) Non-point pollution sources such as storm water runoff in urban areas, poor or nonexistent sanitation facilities in rural areas, and the flushing of livestock manure by rainfall and snowmelt are also causes of contamination (C20.4.5). Pollution from point sources such as mining has had devastating local and regional impacts on the biota of inland waters.

The cell color indicates impact of each driver on biodiversity in each type of ecosystem over the past 50–100 years. High impact means that over the last century the particular driver has significantly altered biodiversity in that biome; low impact indicates that it has had little influence on biodiversity in the biome. The arrows indicate the trend in the driver. Horizontal arrows indicate a continuation of the current level of impact; diagonal and vertical arrows indicate progressively increasing trends in impact. Thus for example, if an ecosystem had experienced a very high impact of a particular driver in the past century (such as the impact of invasive species on islands), a horizontal arrow indicates that this very high impact is likely to continue. This Figure is based on expert opinion consistent with and based on the analysis of drivers of change in the various chapters of the assessment report of the MA Condition and Trends Working Group. The Figure presents global impacts and trends that may be different from those in specific regions.

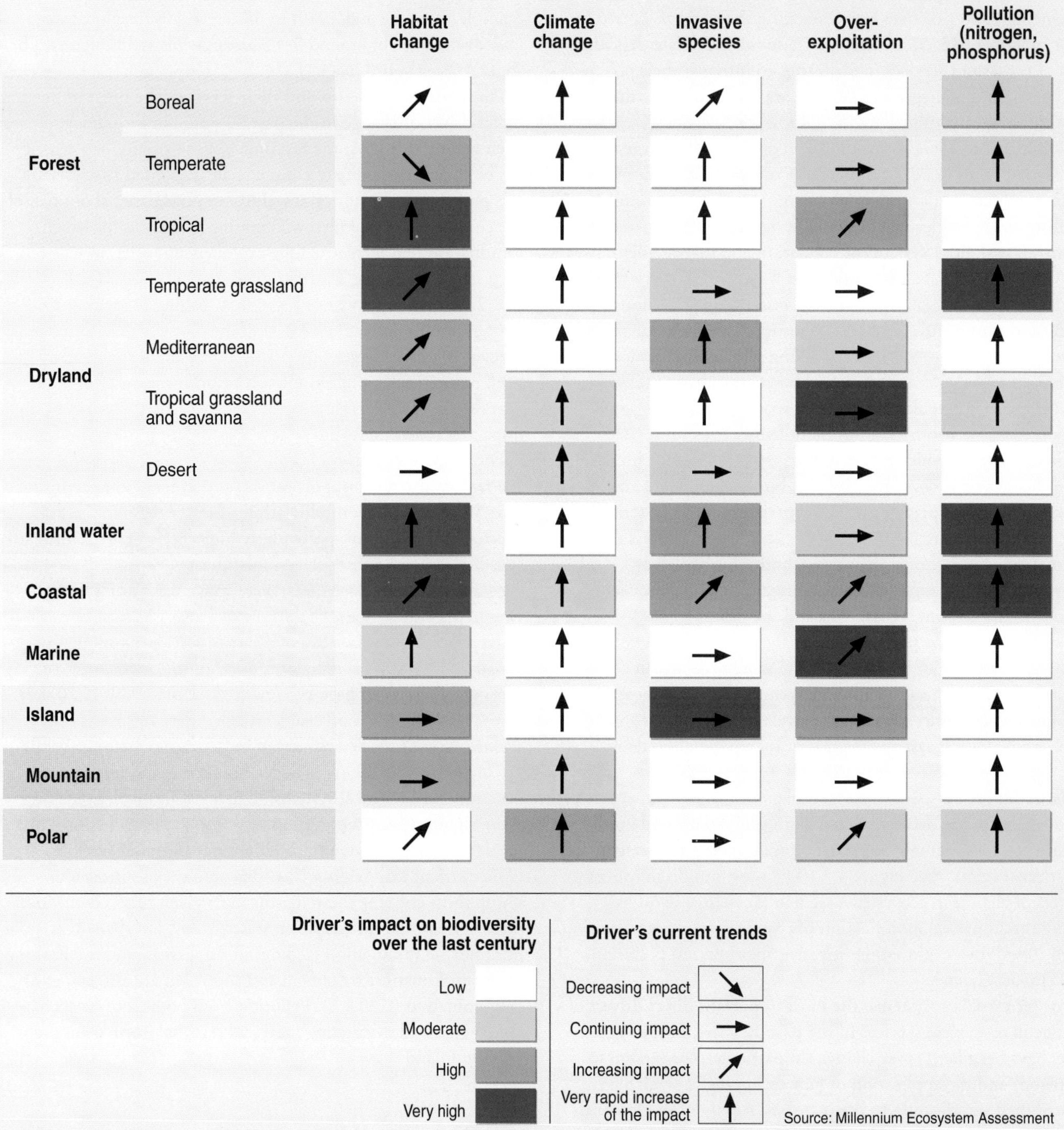

Main Direct Drivers of Change in Biodiversity and Ecosystems. Each cell lists impact (color) and trend (arrow direction).

		Habitat change	Climate change	Invasive species	Over-exploitation	Pollution (nitrogen, phosphorus)
Forest	Boreal	Low / increasing	Low / very rapid	Low / increasing	Moderate / continuing	Moderate / very rapid
	Temperate	High / decreasing	Low / very rapid	Low / very rapid	Moderate / continuing	Moderate / very rapid
	Tropical	Very high / very rapid	Low / very rapid	Low / very rapid	High / increasing	Low / very rapid
Dryland	Temperate grassland	Very high / increasing	Low / very rapid	Moderate / continuing	Low / continuing	Very high / very rapid
	Mediterranean	High / increasing	Low / very rapid	High / very rapid	Moderate / continuing	Low / very rapid
	Tropical grassland and savanna	High / increasing	Low / very rapid	Low / very rapid	Very high / continuing	Moderate / very rapid
	Desert	Low / continuing	Moderate / very rapid	Moderate / continuing	Moderate / continuing	Low / very rapid
Inland water		Very high / very rapid	Low / very rapid	High / very rapid	Moderate / continuing	Very high / very rapid
Coastal		Very high / increasing	Moderate / very rapid	High / increasing	High / increasing	Very high / very rapid
Marine		Moderate / very rapid	Low / very rapid	Low / continuing	Very high / increasing	Low / very rapid
Island		Moderate / continuing	Low / very rapid	Very high / continuing	Moderate / continuing	Low / very rapid
Mountain		High / continuing	Moderate / very rapid	Low / continuing	Low / continuing	Low / very rapid
Polar		Low / increasing	High / very rapid	Low / continuing	Moderate / increasing	Low / very rapid

Driver's impact on biodiversity over the last century

- Low
- Moderate
- High
- Very high

Driver's current trends

- Decreasing impact
- Continuing impact
- Increasing impact
- Very rapid increase of the impact

Source: Millennium Ecosystem Assessment

Coastal ecosystems are affected by multiple direct drivers. Fishing pressures in coastal ecosystems are compounded by a wide array of other drivers, including land-, river-, and ocean-based pollution, habitat loss, invasive species, and nutrient loading. Although human activities have increased sediment flows in rivers by about 20%, reservoirs and water diversions prevent about 30% of sediments from reaching the oceans, resulting in a net reduction of 10% in the sediment delivery to estuaries, which are key nursery areas and fishing grounds (C19.ES). Approximately 17% of the world lives within the boundaries of the MA coastal system (up to an elevation of 50 meters above sea level and no further than 100 kilometers from a coast), and approximately 40% live in the full area within 50 kilometers of a coast. And the absolute number is increasing through a combination of in-migration, high reproduction rates, and tourism (C.SDM). Demand on coastal space for shipping, waste disposal, military and security uses, recreation, and aquaculture is increasing.

The greatest threat to coastal systems is the development-related conversion of coastal habitats such as forests, wetlands, and coral reefs through coastal urban sprawl, resort and port development, aquaculture, and industrialization. Dredging, reclamation and destructive fishing also account for widespread, effectively irreversible destruction. Shore protection structures and engineering works (beach armoring, causeways, bridges, and so on), by changing coastal dynamics, have impacts extending beyond their direct footprints. Nitrogen loading to the coastal zone has increased by about 80% worldwide and has driven coral reef community shifts (C.SDM).

Over the past four decades, excessive nutrient loading has emerged as one of the most important direct drivers of ecosystem change in terrestrial, freshwater, and marine ecosystems. (See Table 4.1.) While the introduction of nutrients into ecosystems can have both beneficial effects (such as increased crop productivity) and adverse effects (such as eutrophication of inland and coastal waters), the beneficial effects will eventually reach a plateau as more nutrients are added (that is, additional inputs will not lead to further increases in crop yield), while the harmful effects will continue to grow.

Synthetic production of nitrogen fertilizer has been an important driver for the remarkable increase in food production that has occurred during the past 50 years (S7.3.2). World consumption of nitrogenous fertilizers grew nearly eightfold between 1960 and 2003, from 10.8 million tons to 85.1 million tons. As much as 50% of the nitrogen fertilizer applied may be lost to the environment, depending on how well the application is managed. Since excessive nutrient loading is largely the result of applying more nutrients than crops can use, it harms both farm incomes and the environment (S7.3.2).

Excessive flows of nitrogen contribute to eutrophication of freshwater and coastal marine ecosystems and acidification of freshwater and terrestrial ecosystems (with implications for biodiversity in these ecosystems). To some degree, nitrogen also plays a

Table 4.1. Increase in Nitrogen Fluxes in Rivers to Coastal Oceans due to Human Activities Relative to Fluxes prior to the Industrial and Agricultural Revolutions (R9 Table 9.1)	
Labrador and Hudson's Bay	no change
Southwestern Europe	3.7-fold
Great Lakes/St. Lawrence basin	4.1-fold
Baltic Sea watersheds	5-fold
Mississippi River basin	5.7-fold
Yellow River basin	10-fold
Northeastern United States	11-fold
North Sea watersheds	15-fold
Republic of Korea	17-fold

role in the creation of ground-level ozone (which leads to loss of agricultural and forest productivity), destruction of ozone in the stratosphere (which leads to depletion of the ozone layer and increased UV-B radiation on Earth, causing increased incidence of skin cancer), and climate change. The resulting health effects include the consequences of ozone pollution on asthma and respiratory function, increased allergies and asthma due to increased pollen production, the risk of blue-baby syndrome, increased risk of cancer and other chronic diseases from nitrates in drinking water, and increased risk of a variety of pulmonary and cardiac diseases from production of fine particles in the atmosphere (R9.ES).

Phosphorus application has increased threefold since 1960, with a steady increase until 1990 followed by a leveling off at a level approximately equal to applications in the 1980s. While phosphorus use has increasingly concentrated on phosphorus-deficient soils, the growing phosphorus accumulation in soils contributes to high levels of phosphorus runoff. As with nitrogen loading, the potential consequences include eutrophication of coastal and freshwater ecosystems, which can lead to degraded habitat for fish and decreased quality of water for consumption by humans and livestock.

Many ecosystem services are reduced when inland waters and coastal ecosystems become eutrophic. Water from lakes that experience algal blooms is more expensive to purify for drinking or other industrial uses. Eutrophication can reduce or eliminate fish populations. Possibly the most apparent loss in services is the loss of many of the cultural services provided by lakes. Foul odors of rotting algae, slime-covered lakes, and toxic chemicals produced by some blue-green algae during blooms keep people from

swimming, boating, and otherwise enjoying the aesthetic value of lakes (S7.3.2).

Climate change in the past century has already had a measurable impact on ecosystems. Earth's climate system has changed since the preindustrial era, in part due to human activities, and it is projected to continue to change throughout the twenty-first century. During the last 100 years, the global mean surface temperature has increased by about 0.6° Celsius, precipitation patterns have changed spatially and temporally, and global average sea level rose by 0.1–0.2 meters (S7.ES). Observed changes in climate, especially warmer regional temperatures, have already affected biological systems in many parts of the world. There have been changes in species distributions, population sizes,

and the timing of reproduction or migration events, as well as an increase in the frequency of pest and disease outbreaks, especially in forested systems. The growing season in Europe has lengthened over the last 30 years (R13.1.3). Although it is not possible to determine whether the extreme temperatures were a result of human-induced climate change, many coral reefs have undergone major, although often partially reversible, bleaching episodes when sea surface temperatures have increased during one month by 0.5–1° Celsius above the average of the hottest months. Extensive coral mortality has occurred with observed local increases in temperature of 3° Celsius (R13.1.3).

5. *How might ecosystems and their services change in the future under various plausible scenarios?*

The MA developed four global scenarios to explore plausible futures for ecosystems and human well-being. (See Box 5.1.) The scenarios were developed with a focus on conditions in 2050, although they include some information through the end of the century. They explored two global development paths, one in which the world becomes increasingly globalized and the other in which it becomes increasingly regionalized, as well as two different approaches to ecosystem management, one in which actions are reactive and most problems are addressed only after they become obvious and the other in which ecosystem management is proactive and policies deliberately seek to maintain ecosystem services for the long term:

■ *Global Orchestration:* This scenario depicts a globally connected society that focuses on global trade and economic liberalization and takes a reactive approach to ecosystem problems but that also takes strong steps to reduce poverty and inequality and to invest in public goods such as infrastructure and education. Economic growth is the highest of the four scenarios, while this scenario is assumed to have the lowest population in 2050.

■ *Order from Strength:* This scenario represents a regionalized and fragmented world that is concerned with security and protection, emphasizes primarily regional markets, pays little attention to public goods, and takes a reactive approach to ecosystem problems. Economic growth rates are the lowest of the scenarios (particularly low in developing countries) and decrease with time, while population growth is the highest.

■ *Adapting Mosaic:* In this scenario, regional watershed-scale ecosystems are the focus of political and economic activity. Local institutions are strengthened and local ecosystem management strategies are common; societies develop a strongly proactive approach to the management of ecosystems. Economic growth rates are somewhat low initially but increase with time, and the population in 2050 is nearly as high as in *Order from Strength*.

■ *TechnoGarden:* This scenario depicts a globally connected world relying strongly on environmentally sound technology, using highly managed, often engineered, ecosystems to deliver ecosystem services, and taking a proactive approach to the management of ecosystems in an effort to avoid problems. Economic growth is relatively high and accelerates, while population in 2050 is in the mid-range of the scenarios.

The scenarios are not predictions; instead, they were developed to explore the unpredictable and uncontrollable features of change in ecosystem services and a number of socioeconomic factors. No scenario represents business as usual, although all begin from current conditions and trends. The future will represent a mix of approaches and consequences described in the scenarios, as well as events and innovations that have not yet been imagined. No scenario is likely to match the future as it actually occurs. These four scenarios were not designed to explore the entire range

of possible futures for ecosystem services—other scenarios could be developed with either more optimistic or more pessimistic outcomes for ecosystems, their services, and human well-being.

The scenarios were developed using both quantitative models and qualitative analysis. For some drivers (such as land use change and carbon emissions) and some ecosystem services (such as water withdrawals and food production), quantitative projections were calculated using established, peer-reviewed global models. Other drivers (such as economic growth and rates of technological change), ecosystem services (particularly supporting and cultural services such as soil formation and recreational opportunities), and human well-being indicators (such as human health and social relations) were estimated qualitatively. In general, the quantitative models used for these scenarios addressed incremental changes but failed to address thresholds, risk of extreme events, or impacts of large, extremely costly, or irreversible changes in ecosystem services. These phenomena were addressed qualitatively, by considering the risks and impacts of large but unpredictable ecosystem changes in each scenario.

(continued on page 74)

Box 5.1. MA Scenarios

Global Orchestration

The *Global Orchestration* scenario depicts a globally connected society in which policy reforms that focus on global trade and economic liberalization are used to reshape economies and governance, emphasizing the creation of markets that allow equitable participation and provide equitable access to goods and services. These policies, in combination with large investments in global public health and the improvement of education worldwide, generally succeed in promoting economic expansion and lifting many people out of poverty into an expanding global middle class. Supranational institutions in this globalized scenario are well placed to deal with global environmental problems such as climate change and fisheries decline. However, the reactive approach to ecosystem management makes people vulnerable to surprises arising from delayed action. While the focus is on improving the well-being of all people, environmental problems that threaten human well-being are only considered after they become apparent.

Growing economies, expansion of education, and growth of the middle class lead to demands for cleaner cities, less pollution, and a more beautiful environment. Rising income levels bring about changes in global consumption patterns, boosting demand for ecosystem services, including agricultural products such as meat, fish, and vegetables. Growing demand for these services leads to declines in other ones, as forests are converted into cropped area and pasture and the services they formerly provided decline. The problems related to increasing food production, such as loss of wildlands, are not apparent to most people who live in urban areas. They therefore receive only limited attention.

Global economic expansion expropriates or degrades many of the ecosystem services poor people once depended on for survival. While economic growth more than compensates for these losses in some regions by increasing the ability to find substitutes for particular ecosystem services, in many other places, it does not. An increasing number of people are affected by the loss of basic ecosystem services essential for human life. While risks seem manageable in some places, in other places there are sudden, unexpected losses as ecosystems cross thresholds and degrade irreversibly. Loss of potable water supplies, crop failures, floods, species invasions, and outbreaks of environmental pathogens increase in frequency. The expansion of abrupt, unpredictable changes in ecosystems, many with harmful effects on increasingly large numbers of people, is the key challenge facing managers of ecosystem services.

Order from Strength

The *Order from Strength* scenario represents a regionalized and fragmented world that is concerned with security and protection, emphasizes primarily regional markets, and pays little attention to common goods. Nations see looking after their own interests as the best defense against economic insecurity, and the movement of goods, people, and information is strongly regulated and policed. The role of government expands as oil companies, water utilities, and other strategic businesses are either nationalized or subjected to more state oversight. Trade is restricted, large amounts of money are invested in security systems, and technological change slows due to restrictions on the flow of goods and information. Regionalization exacerbates global inequality.

Treaties on global climate change, international fisheries, and trade in endangered species are only weakly and haphazardly implemented, resulting in degradation of the global commons. Local problems often go unresolved, but major problems are sometimes handled by rapid disaster relief to at least temporarily resolve the immediate crisis. Many powerful countries cope with local problems by shifting burdens to other, less powerful ones, increasing the gap between rich and poor. In particular, natural resource–intensive industries are moved from wealthier nations to poorer, less powerful ones. Inequality increases considerably within countries as well.

Ecosystem services become more vulnerable, fragile, and variable in *Order from Strength*. For example, parks and reserves exist within fixed boundaries, but climate changes around them, leading to the unintended extirpation of many species. Conditions for crops are often suboptimal, and the ability of societies to import alternative foods is diminished by trade barriers. As a result, there are frequent shortages of food and water, particularly in poor regions. Low levels of trade tend to restrict the number of invasions by exotic species; ecosystems are less resilient, however, and invaders are therefore more often successful when they arrive.

Adapting Mosaic

In the *Adapting Mosaic* scenario, regional watershed-scale ecosystems are the focus of political and economic activity. This scenario sees the rise of local ecosystem management strategies and the strengthening of local institutions. Investments in human and social capital are geared toward improving knowledge about ecosystem functioning and management, which results in a better understanding of resilience, fragility, and local flexibility of ecosystems. There is optimism that we can learn, but humility about preparing for sur-

prises and about our ability to know everything about managing ecosystems.

There is also great variation among nations and regions in styles of governance, including management of ecosystem services. Some regions explore actively adaptive management, investigating alternatives through experimentation. Others use bureaucratically rigid methods to optimize ecosystem performance. Great diversity exists in the outcome of these approaches: some areas thrive, while others develop severe inequality or experience ecological degradation. Initially, trade barriers for goods and products are increased, but barriers for information nearly disappear (for those who are motivated to use them) due to improving communication technologies and rapidly decreasing costs of access to information.

Eventually, the focus on local governance leads to failures in managing the global commons. Problems like climate change, marine fisheries, and pollution grow worse, and global environmental problems intensify. Communities slowly realize that they cannot manage their local areas because global and regional problems are infringing on them, and they begin to develop networks among communities, regions, and even nations to better manage the global commons. Solutions that were effective locally are adopted among networks. These networks of regional successes are especially common in situations where there are mutually beneficial opportunities for coordination, such as along river valleys. Sharing good solutions and discarding poor ones eventually improves approaches to a variety

of social and environmental problems, ranging from urban poverty to agricultural water pollution. As more knowledge is collected from successes and failures, provision of many services improves.

TechnoGarden

The *TechnoGarden* scenario depicts a globally connected world relying strongly on technology and highly managed, often engineered ecosystems to deliver ecosystem services. Overall efficiency of ecosystem service provision improves, but it is shadowed by the risks inherent in large-scale human-made solutions and rigid control of ecosystems. Technology and market-oriented institutional reform are used to achieve solutions to environmental problems. These solutions are designed to benefit both the economy and the environment. These changes co-develop

with the expansion of property rights to ecosystem services, such as requiring people to pay for pollution they create or paying people for providing key ecosystem services through actions such as preservation of key watersheds. Interest in maintaining, and even increasing, the economic value of these property rights, combined with an interest in learning and information, leads to a flowering of ecological engineering approaches for managing ecosystem services. Investment in green technology is accompanied by a significant focus on economic development and education, improving people's lives and helping them understand how ecosystems make their livelihoods possible.

A variety of problems in global agriculture are addressed by focusing on the multifunctional aspects of agriculture and a global reduction of agricultural subsidies and trade barriers. Recognition of the role of agricultural diversification encourages farms to produce a variety of ecological services rather than simply maximizing food production. The combination of these movements stimulates the growth of new markets for ecosystem services, such as tradable nutrient runoff permits, and the development of technology for increasingly sophisticated ecosystem management. Gradually, environmental entrepreneurship expands as new property rights and technologies co-evolve to stimulate the growth of companies and cooperatives providing reliable ecosystem services to cities, towns, and individual property owners.

Innovative capacity expands quickly in developing nations. The reliable provision of ecosystem services as a component of economic growth, together with enhanced uptake of technology due to rising income levels, lifts many of the world's poor into a global middle class. Elements of human well-being associated with social relations decline in this scenario due to great loss of local culture, customs, and traditional knowledge and the weakening of civil society institutions as an increasing share of interactions take place over the Internet. While the provision of basic ecosystem services improves the well-being of the world's poor, the reliability of the services, especially in urban areas, become more critical and is increasingly difficult to ensure. Not every problem has succumbed to technological innovation. Reliance on technological solutions sometimes creates new problems and vulnerabilities. In some cases, societies seem to be barely ahead of the next threat to ecosystem services. In such cases new problems often seem to emerge from the last solution, and the costs of managing the environment are continually rising. Environmental breakdowns that affect large numbers of people become more common. Sometimes new problems seem to emerge faster than solutions. The challenge for the future is to learn how to organize socioecological systems so that ecosystem services are maintained without taxing society's ability to implement solutions to novel, emergent problems.

Projected Changes in Indirect and Direct Drivers under MA Scenarios

In the four MA scenarios, during the first half of the twenty-first century the array of both indirect and direct drivers affecting ecosystems and their services is projected to remain largely the same as over the last half-century, but the relative importance of different drivers will begin to change. Some factors (such as global population growth) will begin to decline in importance and others (distribution of people, climate change, and changes to nutrient cycles) will gain more importance. (See Tables 5.1, 5.2, and 5.3.)

Statements of certainty associated with findings related to the MA scenarios are conditional statements; they refer to level of certainty or uncertainty in the particular projection should that scenario and its associated changes in drivers unfold. They do not indicate the likelihood that any particular scenario and its associated projection will come to pass. With that caveat in mind, the four MA scenarios describe these changes between 2000 and 2050 (or in some cases 2100):

■ *Population is projected to grow to 8.1–9.6 billion in 2050 (medium to high certainty) and to 6.8–10.5 billion in 2100, depending on the scenario* (S7.2.1). (See Figure 5.1.) The rate of global population growth has already peaked, at 2.1% per year in the late 1960s, and had fallen to 1.35% per year in 2000, when global population reached 6 billion (S7.ES). Population growth over the next several decades is expected to be concentrated in the poorest, urban communities in sub-Saharan Africa, South Asia, and the Middle East (S7.ES).

■ *Per capita income is projected to increase two- to fourfold, depending on the scenario (low to medium certainty)* (S7.2.2). Gross world product is projected to increase roughly three to sixfold in the different scenarios. Increasing income leads to increasing per capita consumption in most parts of the world for most resources and it changes the structure of consumption. For example, diets tend to become higher in animal protein as income rises.

■ *Land use change (primarily the continuing expansion of agriculture) is projected to continue to be a major direct driver of change in terrestrial and freshwater ecosystems (medium to high certainty)* (S9.ES). At the global level and across all scenarios, land use change is projected to remain the dominant driver of biodiversity change in terrestrial ecosystems, consistent with the pattern over the past 50 years, followed by changes in climate and nitrogen deposition (S10.ES). However, other direct drivers may be more important than land use change in particular biomes. For example, climate change is likely to be the dominant driver of biodiversity change in tundra and deserts. Species invasions and water extraction are important drivers for freshwater ecosystems.

■ *Nutrient loading is projected to become an increasingly severe problem, particularly in developing countries.* Nutrient loading already has major adverse effects on freshwater ecosystems and coastal regions in both industrial and developing countries. These impacts include toxic algae blooms, other human health problems, fish kills, and damage to habitats such as coral reefs. Three out of the four MA scenarios project that the global flux of nitrogen to coastal ecosystems will increase by 10–20% by 2030 (*medium certainty*) (S9.3.7.2). (See Figure 5.2.) River nitrogen will not change in most industrial countries, while a 20–30% increase is projected for developing countries, particularly in Asia.

■ *Climate change and its impacts (such as sea level rise) are projected to have an increasing effect on biodiversity and ecosystem services (medium certainty)* (S9.ES). Under the four MA scenarios, global temperature is expected to increase significantly—1.5–2.0° Celsius above preindustrial level in 2050 and 2.0–3.5° Celsius above it in 2100, depending on the scenario and using

(*continued on page 78*)

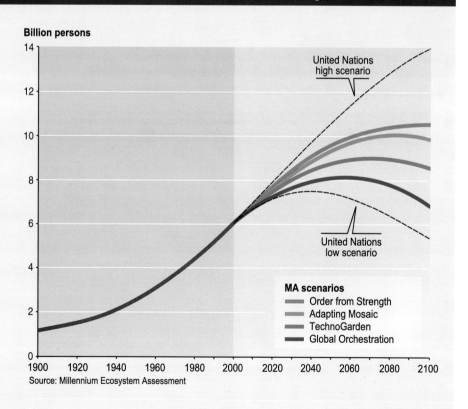

Figure 5.1. MA WORLD POPULATION SCENARIOS (S7 Fig 7.2)

Billion persons

United Nations high scenario

United Nations low scenario

MA scenarios
- Order from Strength
- Adapting Mosaic
- TechnoGarden
- Global Orchestration

Source: Millennium Ecosystem Assessment

	Global Orchestration	Order from Strength		Adapting Mosaic	TechnoGarden
		Industrial Countries[a]	Developing Countries[a]		
Indirect Drivers					
Demographics	high migration; low fertility and mortality levels 2050 population: 8.1 billion	high fertility and mortality levels (especially in developing countries); low migration 2050 population: 9.6 billion		high fertility level; high mortality levels until 2010 then medium by 2050; low migration 2050 population: 9.5 billion	medium fertility and mortality levels; medium migration 2050 population: 8.8 billion
Average income growth	high	medium	low	similar to *Order from Strength* but with increasing growth rates toward 2050	lower than *Global Orchestration*, but catching up toward 2050
GDP growth rates/capita per year until 2050	Global: 1995–2020: 2.4% per year 2020–2050: 3.0% per year industrialized c.: 1995–2020: 2.5% per year 2020–2050: 2.1% per year developing c.: 1995–2020: 3.8% per year 2020–2050: 4.8% per year	1995–2020: 1.4% per year 2020–2050: 1.0% per year 1995–2020: 2.1% per year 2020–2050: 1.4% per year	 1995–2020: 2.4% per year 2020–2050: 2.3% per year	1995–2020: 1.5% per year 2020–2050: 1.9% per year industrialized c.: 1995-2020: 2.0% per year 2020–2050: 1.7% per year developing c.: 1995–2020: 2.8% per year 2020–2050: 3.5% per year	1995–2020: 1.9% per year 2020–2050: 2.5% per year industrialized c.: 1995–2020: 2.3% per year 2020–2050: 1.9% per year developing c.: 1995–2020: 3.2% per year 2020–2050: 4.3% per year
Income distribution	becomes more equal	similar to today		similar to today, then becomes more equal	becomes more equal
Investments into new produced assets	high	medium	low	begins like *Order from Strength*, then increases in tempo	high
Investments into human capital	high	medium	low	begins like *Order from Strength*, then increases in tempo	medium
Overall trend in technology advances	high	low		medium-low	medium in general; high for environmental technology
International cooperation	strong	weak – international competition		weak – focus on local environment	strong
Attitude toward environmental policies	reactive	reactive		proactive – learning	proactive

(continued on page 76)

Table 5.1. MAIN ASSUMPTIONS CONCERNING INDIRECT AND DIRECT DRIVING FORCES USED IN THE MA SCENARIOS (S.SDM)

	Global Orchestration	Order from Strength		Adapting Mosaic	TechnoGarden
		Industrial Countries[a]	Developing Countries[a]		
Indirect Drivers *(continued)*					
Energy demand and lifestyle	Energy-intensive	regionalized assumptions		regionalized assumptions	high level of energy efficiency; saturation in energy use
Energy supply	market liberalization; selects least-cost options; rapid technology change	focus on domestic energy resources		some preference for clean energy resources	preference for renewable energy resources and rapid technology change
Climate policy	no	no		no	yes, aims at stabilization of CO_2-equivalent concentration at 550 ppmv
Approach to achieving sustainability	economic growth leads to sustainable development	national-level policies; conservation; reserves, parks		local-regional co-management; common-property institutions	green-technology; eco-efficiency; tradable ecological property rights
Direct Drivers					
Land use change	global forest loss until 2025 slightly below historic rate, stabilizes after 2025; ~10% increase in arable land	global forest loss faster than historic rate until 2025; near current rate after 2025; ~20% increase in arable land compared with 2000		global forest loss until 2025 slightly below historic rate, stabilizes after 2025; ~10% increase in arable land	net increase in forest cover globally until 2025; slow loss after 2025; ~9% increase in arable land
Greenhouse gas emissions by 2050	CO_2: 20.1 GtC-eq CH$_4$: 3.7 GtC-eq N$_2$O: 1.1 GtC-eq other GHG: 0.7 GtC-eq	CO_2: 15.4 GtC-eq CH$_4$: 3.3 GtC-eq N$_2$O: 1.1 GtC-eq other GHG: 0.5 GtC-eq		CO_2: 13.3 GtC-eq CH$_4$: 3.2 GtC-eq N$_2$O: 0.9 GtC-eq other GHG: 0.6 GtC-eq	CO_2: 4.7 GtC-eq CH$_4$: 1.6 GtC-eq N$_2$O: 0.6 GtC-eq other GHG: 0.2 GtC-eq
Air pollution emissions	SO_2 emissions stabilize; NOx emissions increase from 2000 to 2050	both SO_2 and NOx emissions increase globally		SO_2 emissions decline; NOx emissions increase slowly	strong reductions in SO_2 and NOx emissions
Climate change	2.0°C in 2050 and 3.5°C in 2100 above preindustrial	1.7°C in 2050 and 3.3°C in 2100 above preindustrial		1.9°C in 2050 and 2.8°C in 2100 above preindustrial	1.5°C in 2050 and 1.9°C in 2100 above preindustrial
Nutrient loading	increase in N transport in rivers	increase in N transport in rivers		increase in N transport in rivers	decrease in N transport in rivers

[a] These categories refer to the countries at the beginning of the scenario; some countries may change categories during the course of the 50 years.

Table 5.2. Outcomes of Scenarios for Ecosystem Services in 2050 Compared with 2000 (S.SDM)

Definitions of "enhanced" and "degraded" are provided the note below.

	Global Orchestration		Order from Strength		Adapting Mosaic		TechnoGarden	
Provisioning Services	Industrial Countries[a]	Developing Countries[a]	Industrial Countries[a]	Developing Countries[a]	Industrial Countries[a]	Developing Countries[a]	Industrial Countries[a]	Developing Countries[a]
Food (extent to which demand is met)	▲	▲	◀▶	▼	◀▶	▼	▲	▲
Fuel	▲	▲	▲	▲	▲	▲	▲	▲
Genetic resources	◀▶	◀▶	▼	▼	▲	▲	◀▶	▲
Biochemicals/ pharmaceutical discoveries	▼	▲	▼	▼	◀▶	◀▶	▲	▲
Ornamental resources	◀▶	◀▶	◀▶	▼	▲	▲	◀▶	◀▶
Fresh water	▲	▲	◀▶	▼	▲	▼	▲	◀▶
Regulating Services								
Air quality regulation	◀▶	◀▶	◀▶	▼	◀▶	◀▶	▲	▲
Climate regulation	◀▶	◀▶	▼	▼	◀▶	◀▶	▲	▲
Water regulation	◀▶	▼	▼	▼	▲	▲	◀▶	▲
Erosion control	◀▶	▼	▼	▼	▲	▲	◀▶	▲
Water purification	◀▶	▼	▼	▼	▲	▲	◀▶	▲
Disease control: human	◀▶	▲	◀▶	▼	◀▶	▲	▲	▲
Disease control: pests	◀▶	▼	▼	▼	▲	▲	◀▶	◀▶
Pollination	▼	▼	▼	▼	◀▶	◀▶	▼	▼
Storm protection	◀▶	▼	◀▶	▼	▲	▲	▲	◀▶
Cultural Services								
Spiritual/religious values	◀▶	◀▶	◀▶	▼	▲	▲	▼	▼
Aesthetic values	◀▶	◀▶	◀▶	▼	▲	▲	◀▶	◀▶
Recreation and ecotourism	▼	▲	▼	▲	▼	▼	▲	▲
Cultural diversity	▼	▼	▼	▼	▲	▲	▼	▼
Knowledge systems (diversity and memory)	◀▶	▼	▼	▼	▲	▲	◀▶	◀▶

Legend: ▲ = increase, ◀▶ = remains the same as in 2000, ▼ = decrease

Note: For provisioning services, we define enhancement to mean increased production of the service through changes in area over which the service is provided (e.g., spread of agriculture) or increased production per unit area. We judge the production to be degraded if the current use exceeds sustainable levels. For regulating services, enhancement refers to a change in the service that leads to greater benefits for people (e.g., the service of disease regulation could be improved by eradication of a vector known to transmit a disease to people). Degradation of regulating services means a reduction in the benefits obtained from the service, either through a change in the service (e.g., mangrove loss reducing the storm protection benefits of an ecosystem) or through human pressures on the service exceeding its limits (e.g., excessive pollution exceeding the capability of ecosystems to maintain water quality). For cultural services, degradation refers to a change in the ecosystem features that decreases the cultural (recreational, aesthetic, spiritual, etc.) benefits provided by the ecosystem, while enhancement refers to a change that increases them.

[a] These categories refer to the countries at the beginning of the scenario; some countries may change categories during the course of the 50 years.

Table 5.3. Outcomes of Scenarios for Human Well-being in 2050 Compared with 2000

Services	Global Orchestration		Order from Strength		Adapting Mosaic		TechnoGarden	
	Industrial Countries[a]	Developing Countries[a]	Industrial Countries[a]	Developing Countries[a]	Industrial Countries[a]	Developing Countries[a]	Industrial Countries[a]	Developing Countries[a]
Material well-being	▲	▲	▲	▼	◀▶	▲	▲	▲
Health	▲	▲	▲	▼	▲	▲	▲	▲
Security	▲	▲	▼	▼	▲	▲	▲	▲
Social relations	◀▶	▲	▼	▼	▲	▲	▼	▼
Freedom and choice	◀▶	▲	▼	▼	▲	▲	▲	▲

Legend: ▲ = increase, ◀▶ = remains the same as in 2000, ▼ = decrease

[a] These categories refer to the countries at the beginning of the scenario; some countries may change categores during the course of 50 years.

Figure 5.2. Comparison of Global River Nitrogen Export from Natural Ecosystems, Agricultural Systems, and Sewage Effluents, 1975 and 1990, with Model Results for the MA Scenarios in 2030 (S9 Fig 9.21)

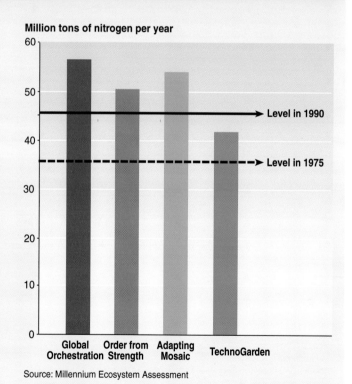

Million tons of nitrogen per year

Level in 1990
Level in 1975

Global Orchestration   Order from Strength   Adapting Mosaic   TechnoGarden

Source: Millennium Ecosystem Assessment

a median estimate for climate sensitivity (2.5°C for a doubling of the CO_2 concentration) (*medium certainty*). The IPCC reported a range of temperature increase for the scenarios used in the Third Assessment Report of 2.0–6.4° Celsius compared with pre-industrial levels, with about half of this range attributable to the differences in scenarios and the other half to differences in climate models. The smaller, somewhat lower, range of the MA scenarios is thus partly a result of using only one climate model (and one estimate of climate sensitivity) but also the result of including climate policy responses in some scenarios as well as differences in assumptions for economic and population growth. The scenarios project an increase in global average precipitation (*medium certainty*), but some areas will become more arid while others will become more moist. Climate change will directly alter ecosystem services, for example, by causing changes in the productivity and growing zones of cultivated and noncultivated vegetation. It is also projected to change the frequency of extreme events, with associated risks to ecosystem services. Finally, it is projected to indirectly affect ecosystem services in many ways, such as by causing sea level to rise, which threatens mangroves and other vegetation that now protect shorelines.

Climate change is projected to further adversely affect key development challenges, including providing clean water, energy services, and food; maintaining a healthy environment; and conserving ecological systems, their biodiversity, and their associated ecological goods and services (R13.1.3).

- Climate change is projected to exacerbate the loss of biodiversity and increase the risk of extinction for many species, especially those already at risk due to factors such as low population numbers, restricted or patchy habitats, and limited climatic ranges (*medium to high certainty*).
- Water availability and quality are projected to decrease in many arid and semiarid regions (*high certainty*).
- The risk of floods and droughts is projected to increase (*high certainty*).

- Sea level is projected to rise by 8–88 centimeters.
- The reliability of hydropower and biomass production is projected to decrease in some regions (*high certainty*).
- The incidence of vector-borne diseases such as malaria and dengue and of waterborne diseases such as cholera is projected to increase in many regions (*medium to high certainty*), and so too are heat stress mortality and threats of decreased nutrition in other regions, along with severe weather traumatic injury and death (*high certainty*).
- Agricultural productivity is projected to decrease in the tropics and sub-tropics for almost any amount of warming (*low to medium certainty*), and there are projected adverse effects on fisheries.
- Projected changes in climate during the twenty-first century are very likely to be without precedent during at least the past 10,000 years and, combined with land use change and the spread of exotic or alien species, are likely to limit both the capability of species to migrate and the ability of species to persist in fragmented habitats.

■ *By the end of the century, climate change and its impacts may be the dominant direct drivers of biodiversity loss and the change in ecosystem services globally* (R13). Harm to biodiversity will grow with both increasing rates in change in climate and increasing absolute amounts of change. For ecosystem services, some services in some regions may initially benefit from increases in temperature or precipitation expected under climate scenarios, but the balance of evidence suggests that there will be a significant net harmful impact on ecosystem services worldwide if global mean surface temperature increases more than 2° Celsius above preindustrial levels or at rates greater than 0.2° Celsius per decade (*medium certainty*). There is a wide band of uncertainty in the amount of warming that would result from any stabilized greenhouse gas concentration, but based on IPCC projections this would require an eventual CO_2 stabilization level of less than 450 parts per million carbon dioxide (*medium certainty*).

This judgment is based on the evidence that an increase of about 2° Celsius above preindustrial levels in global mean surface temperature would represent a transition between the negative effects of climate change being felt in only some regions of the world to most regions of the world. For example, below an increase of about 2° Celsius, agricultural productivity is projected to be adversely affected in the tropics and sub-tropics, but beneficially affected in most temperate and high-latitude regions, whereas more warming than that is projected to have adverse impacts on agricultural productivity in many temperate regions. A 2° increase would have both positive and negative economic impacts, but most people would be adversely affected—that is, there would be predominantly negative economic effects. It would pose a risk to many unique and threatened ecological systems and lead to the extinction of numerous species. And it would lead to a significant increase in extreme climatic events and adversely affect water resources in countries that are already water-scarce or water-stressed and would affect human health and property.

Changes in Ecosystems

Rapid conversion of ecosystems is projected to continue under all MA scenarios in the first half of the twenty-first century. Roughly 10–20% (*low to medium certainty*) of current grassland and forestland is projected to be converted to other uses between now and 2050, mainly due to the expansion of agriculture and, secondarily, because of the expansion of cities and infrastructure (S9.ES). The biomes projected to lose habitat and local species at the fastest rate in the next 50 years are warm mixed forests, savannas, scrub, tropical forests, and tropical woodlands (S10.ES). Rates of conversion of ecosystems are highly dependent on future development scenarios and in particular on changes in population, wealth, trade, and technology.

Habitat loss in terrestrial environments is projected to accelerate decline in local diversity of native species in all four scenarios by 2050 (*high certainty*) (S.SDM). Loss of habitat results in the immediate extirpation of local populations and the loss of the services that these populations provided.

The habitat losses projected in the MA scenarios will lead to global extinctions as numbers of species approach equilibrium with the remnant habitat (*high certainty*) (S.SDM, S10.ES). The equilibrium number of plant species is projected to be reduced by roughly 10–15% as a result of habitat loss from 1970 to 2050 in the MA scenarios (*low certainty*). Other terrestrial taxonomic groups are likely to be affected to a similar extent. The pattern of extinction through time cannot be estimated with any precision, because some species will be lost immediately when their habitat is modified but others may persist for decades or centuries. Time lags between habitat reduction and extinction provide an opportunity for humans to deploy restoration practices that may rescue those species that otherwise may be in a trajectory toward extinction. Significant declines in freshwater fish species diversity are also projected due to the combined effects of climate change, water withdrawals, eutrophication, acidification, and increased invasions by nonindigenous species (*low certainty*). Rivers that are expected to lose fish species are concentrated in poor tropical and sub-tropical countries.

Changes in Ecosystem Services and Human Well-being

In three of the four MA scenarios, ecosystem services show net improvements in at least one of the three categories of provisioning, regulating, and cultural services (S.SDM). These three categories of ecosystem services are all in worse condition in 2050 than they are today in only one MA scenario—*Order from Strength*. (See Figure 5.3.) However, even in scenarios showing improvement in one or more categories of ecosystem services, biodiversity loss continues at high rates.

The Figure shows the net change in the number of ecosystem services enhanced or degraded in the MA scenarios in each category of services for industrial and developing countries expressed as a percentage of the total number of services evaluated in that category. Thus, 100% degradation means that all the services in the category were degraded in 2050 compared with 2000, while 50% improvement could mean that three out of six services were enhanced and the rest were unchanged or that four out of six were enhanced and one was degraded. The total number of services evaluated for each category was six provisioning services, nine regulating services, and five cultural services.

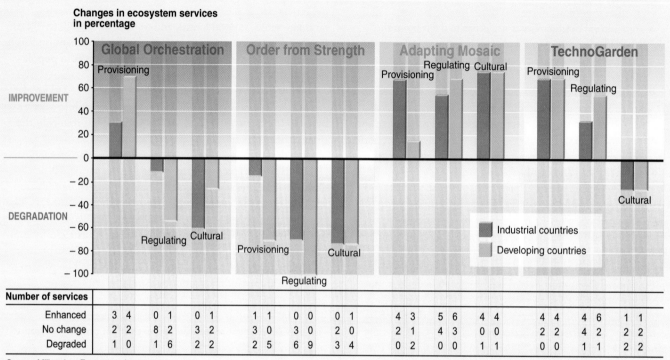

Number of services	Global Orchestration						Order from Strength						Adapting Mosaic						TechnoGarden					
Enhanced	3	4	0	1	0	1	1	1	0	0	0	1	4	3	5	6	4	4	4	4	4	6	1	1
No change	2	2	8	2	3	2	3	0	3	0	2	0	2	1	4	3	0	0	2	2	4	2	2	2
Degraded	1	0	1	6	2	2	2	5	6	9	3	4	0	2	0	0	1	1	0	0	1	1	2	2

Source: Millennium Ecosystem Assessment

The following changes to ecosystem services and human well-being were common to all four MA scenarios and thus may be likely under a wide range of plausible futures (S.SDM):

■ *Human use of ecosystem services increases substantially under all MA scenarios during the next 50 years.* In many cases this is accompanied by degradation in the quality of the service and sometimes, in cases where the service is being used unsustainably, a reduction in the quantity of the service available. (See Appendix A.) The combination of growing populations and growing per capita consumption increases the demand for ecosystem services, including water and food. For example, demand for food crops (measured in tons) is projected to grow by 70–85% by 2050 (S9.4.1) and global water withdrawals increase by 20–85% across the MA scenarios (S9 Fig 9.35). Water withdrawals are projected to increase signifi-

cantly in developing countries but to decline in OECD countries (*medium certainty*) (S.SDM). In some cases, this growth in demand will be met by unsustainable uses of the services, such as through continued depletion of marine fisheries. Demand is dampened somewhat by increasing efficiency in use of resources. The quantity and quality of ecosystem services will change dramatically in the next 50 years as productivity of some services is increased to meet demand, as humans use a greater fraction of some services, and as some services are diminished or degraded. Ecosystem services that are projected to be further impaired by ecosystem change include fisheries, food production in drylands, quality of fresh waters, and cultural services.

■ *Food security is likely to remain out of reach for many people.* Child malnutrition will be difficult to eradicate even by 2050 (*low to medium certainty*) and is projected to increase in some regions in some MA scenarios, despite increasing food supply under all four scenarios (*medium to high certainty*) and more

diversified diets in poor countries (*low to medium certainty*) (S.SDM). Three of the MA scenarios project reductions in child undernourishment by 2050 of between 10% and 60%, but undernourishment increases by 10% in *Order from Strength* (*low certainty*) (S9.4.1). (See Figure 5.4.) This is due to a combination of factors related to food supply systems (inadequate investments in food production and its supporting infrastructure resulting in low productivity increases, varying trade regimes) and food demand and accessibility (continuing poverty in combination with high population growth rates, lack of food infrastructure investments).

■ *Vast, complex changes with great geographic variability are projected to occur in world freshwater resources and hence in their provisioning of ecosystem services in all scenarios* (S.SDM). Climate change will lead to increased precipitation over more than half of Earth's surface, and this will make more water available to society and ecosystems (*medium certainty*). However, increased precipitation is also likely to increase the frequency of flooding in many areas (*high certainty*). Increases in precipitation will not be universal, and climate change will also cause a substantial decrease in precipitation in some areas, with an accompanying decrease in water availability (*medium certainty*). These areas could include highly populated arid regions such as the Middle East and Southern Europe (*low to medium certainty*). While water withdrawals decrease in most industrial countries, they are expected to increase substantially in Africa and some other developing regions, along with wastewater discharges, overshadowing the possible benefits of increased water availability (*medium certainty*).

■ *A deterioration of the services provided by freshwater resources (such as aquatic habitat, fish production, and water supply for households, industry, and agriculture) is expected in developing countries under the scenarios that are reactive to environmental problems* (S9.ES). Less severe but still important declines are expected in the scenarios that are more proactive about environmental problems (*medium certainty*).

■ *Growing demand for fish and fish products leads to an increasing risk of a major and long-lasting collapse of regional marine fisheries* (*low to medium certainty*) (S.SDM). Aquaculture may relieve some of this pressure by providing for an increasing fraction of fish demand. However, this would require aquaculture to reduce its current reliance on marine fish as a feed source.

The future contribution of terrestrial ecosystems to the regulation of climate is uncertain (S9.ES). Carbon release or uptake by ecosystems affects the CO_2 and CH_4 content of the atmosphere at the global scale and thereby affects global climate. Currently, the biosphere is a net sink of carbon, absorbing about 1–2 gigatons a year, or approximately 20% of fossil fuel emissions. It is very likely that the future of this service will be greatly affected by expected land use change. In addition, a higher atmospheric CO_2 concentration is expected to enhance net productivity, but this does not necessarily lead to an increase in the carbon

sink. The limited understanding of soil respiration processes generates uncertainty about the future of the carbon sink. There is *medium certainty* that climate change will increase terrestrial fluxes of CO_2 and CH_4 in some regions (such as in Arctic tundra).

Dryland ecosystems are particularly vulnerable to changes over the next 50 years. The combination of low current levels of human well-being (high rates of poverty, low per capita GDP, high infant mortality rates), a large and growing population, high variability of environmental conditions in dryland regions, and high sensitivity of people to changes in ecosystem services means that continuing land degradation could have profoundly negative impacts on the well-being of a large number of people in these regions (S.SDM). Subsidies of food and water to people in vulnerable drylands can have the unintended effect of increasing the risk of even larger breakdowns of ecosystem services in future years. Local adaptation and conservation practices can mitigate some losses of dryland ecosystem services, although it will be difficult to reverse trends toward loss of food production capacity, water supplies, and biodiversity in drylands.

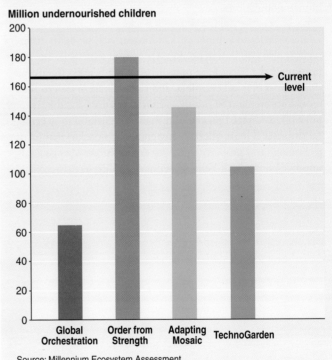

Figure 5.4. NUMBER OF UNDERNOURISHED CHILDREN PROJECTED IN 2050 UNDER MA SCENARIOS

Source: Millennium Ecosystem Assessment

While human health improves under most MA scenarios, under one plausible future health and social conditions in the North and South could diverge (S11). In the more promising scenarios related to health, the number of undernourished children is reduced, the burden of epidemic diseases such as HIV/AIDS, malaria, and tuberculosis would be lowered, improved vaccine development and distribution could allow populations to cope comparatively well with the next influenza pandemic, and the impact of other new diseases such as SARS would also be limited by well-coordinated public health measures.

Under the *Order from Strength* scenario, however, it is plausible that the health and social conditions for the North and South could diverge as inequality increases and as commerce and scientific exchanges between industrial and developing countries decrease. In this case, health in developing countries could become worse, causing a negative spiral of poverty, declining health, and degraded ecosystems. The increased population in the South, combined with static or deteriorating nutrition, could force increased contact between humans and nonagricultural ecosystems, especially to obtain bushmeat and other forest goods. This could lead to more outbreaks of hemorrhagic fever and zoonoses. It is possible, though with low probability, that a more chronic disease could cross from a nondomesticated animal species into humans, at first slowly but then more rapidly colonizing human populations.

Each scenario yields a different package of gains, losses, and vulnerabilities to components of human well-being in different regions and populations (S.SDM). Actions that focus on improving the lives of the poor by reducing barriers to international flows of goods, services, and capital tend to lead to the most improvement in health and social relations for the currently most disadvantaged people. But human vulnerability to ecological surprises is high. Globally integrated approaches that focus on technology and property rights for ecosystem services generally improve human well-being in terms of health, security, social

relations, and material needs. If the same technologies are used globally, however, local culture can be lost or undervalued. High levels of trade lead to more rapid spread of emergent diseases, somewhat reducing the gains in health in all areas. Locally focused, learning-based approaches lead to the largest improvements in social relations.

Order from Strength, which focuses on reactive policies in a regionalized world, has the least favorable outcomes for human well-being, as the global distribution of ecosystem services and human resources that underpin human well-being are increasingly skewed. (See Figure 5.5.) Wealthy populations generally meet most material needs but experience psychological unease. Anxiety, depression, obesity, and diabetes have a greater impact

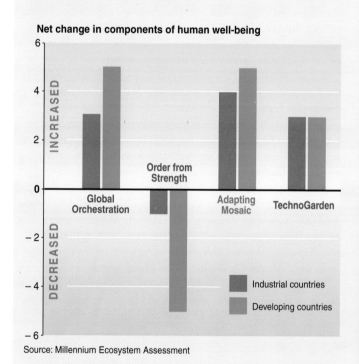

Figure 5.5. Net Change in Components of Human Well-being between 2000 and 2050 under MA Scenarios (Data from Table 5.3)

The Figure shows the number of components of human well-being enhanced minus the number degraded for each scenario between 2000 and 2050 for industrial and developing countries. This qualitative assessment of status examined five components of human well-being: material well-being, health, security, good social relations, and freedom of choice and action.

Net change in components of human well-being

Source: Millennium Ecosystem Assessment

on otherwise privileged populations in this scenario. Disease creates a heavy burden for disadvantaged populations.

Proactive or anticipatory management of ecosystems is generally advantageous in the MA scenarios, but it is particularly beneficial under conditions of changing or novel conditions (S.SDM). (See Table 5.4.) Ecological surprises are inevitable because of the complexity of the interactions and because of limitations in current understanding of the dynamic properties of ecosystems. Currently well understood phenomena that were surprises of the past century include the ability of pests to evolve resistance to biocides, the contribution to desertification of certain types of land use, biomagnification of toxins, and the increase in vulnerability of ecosystem to eutrophication and unwanted species due to removal of predators. While we do not know which surprises lie ahead in the next 50 years, we can be certain that there will be some.

In general, proactive action to manage systems sustainably and to build resilience into systems will be advantageous, particularly when conditions are changing rapidly, when surprise events are likely, or when uncertainty is high. This approach is beneficial largely because the restoration of ecosystems or ecosystem services following their degradation or collapse is generally more costly and time-consuming than preventing degradation, if that is possible at all. Nevertheless, there are costs and benefits to both proactive and reactive approaches, as Table 5.4 indicated.

Table 5.4. Costs and Benefits of Proactive as Contrasted with Reactive Ecosystem Management as Revealed in the MA Scenarios (S.SDM)

	Proactive Ecosystem Management	Reactive Ecosystem Management
Payoffs	benefit from lower risk of unexpected losses of ecosystem services, achieved through investment in more efficient use of resources (water, energy, fertilizer, etc.); more innovation of green technology; capacity to absorb unexpected fluctuations in ecosystem services; adaptable management systems; and ecosystems that are resilient and self-maintaining	avoid paying for monitoring effort
	do well under changing or novel conditions	do well under smoothly or incrementally changing conditions
	build natural, social, and human capital	build manufactured, social, and human capital
Costs	technological solutions can create new problems	expensive unexpected events
	costs of unsuccessful experiments	persistent ignorance (repeating the same mistakes)
	costs of monitoring	lost option values
	some short-term benefits are traded for long-term benefits	inertia of less flexible and adaptable management of infrastructure and ecosystems
		loss of natural capital

6. *What can be learned about the consequences of ecosystem change for human well-being at sub-global scales?*

The MA included a sub-global assessment component to assess differences in the importance of ecosystem services for human well-being around the world (SG.SDM). The Sub-global Working Group included 33 assessments around the world. (See Figure 6.1.) These were designed to consider the importance of ecosystem services for human well-being at local, national, and regional scales. The areas covered in these assessments range from small villages in India and cities like Stockholm and São Paulo to whole countries like Portugal and large regions like southern Africa. In a few cases, the sub-global assessments were designed to cover multiple nested scales. For example, the Southern Africa

study included assessments of the entire region of Africa south of the equator, of the Gariep and Zambezi river basins in that region, and of local communities within those basins. This nested design was included as part of the overall design of the MA to analyze the importance of scale on ecosystem services and human well-being and to study cross-scale interactions. Most assessments, however, were conducted with a focus on the needs of users at a single spatial scale—a particular community, watershed, or region.

The scale at which an assessment is undertaken significantly influences the problem definition and the assessment results (SG.SDM). Findings of assessments done at different scales varied due to the specific questions posed or the information analyzed. Local communities are influenced by global, regional, and local factors. Global factors include commodity prices (global trade asymmetries that influence local production patterns, for instance) and global climate change (such as sea level rise). Regional factors include water supply regimes (safe piped water in rural areas), regional climate (desertification), and geomorphological processes (soil erosion and degradation). Local factors include market access (distance to market), disease prevalence (malaria, for example), or localized climate variability (patchy thunderstorms). Assessments conducted at different scales tended to focus on drivers and impacts most relevant at each scale, yielding different but complementary findings. This provides some of the benefit of a multiscale assessment process, since each component assessment provides a different perspective on the issues addressed.

Although there is overall congruence in the results from global and sub-global assessments for services like water and biodiversity, there are examples where local assessments showed the condition was either better or worse than expected from the global assessment (SG.SDM). For example, the condition of water resources was significantly worse than expected in places like São Paulo and the Laguna Lake Basin in the Philippines. There were more mismatches for biodiversity than for water provisioning because the concepts and measures of biodiversity were more diverse in the sub-global assessments.

Drivers of change act in very distinct ways in different regions (SG7.ES). Though similar drivers might be present in various assessments, their interactions—and thus the processes leading to ecosystem change—differed significantly from one assessment to another. For example, although the Amazon, Central Africa, and Southeast Asia in the Tropical Forest Margins assessment have the same set of individual drivers of land use change (deforestation, road construction, and pasture creation), the interactions among these drivers leading to change differ. Deforestation driven by swidden agriculture is more widespread in upland and foothill zones of Southeast Asia than in other regions. Road

Figure 6.1. MA Sub-global Assessments

Eighteen assessments were approved as components of the MA. Any institution or country was able to undertake an assessment as part of the MA if it agreed to use the MA Conceptual Framework, to centrally involve the intended users as stakeholders and partners, and to meet a set of procedural requirements related to peer review, metadata, transparency, and intellectual property rights. The MA assessments were largely self-funded, although planning grants and some core grants were provided to support some assessments. The MA also drew on information from 15 other sub-global assessments affiliated with the MA that met a subset of these criteria or were at earlier stages in development.

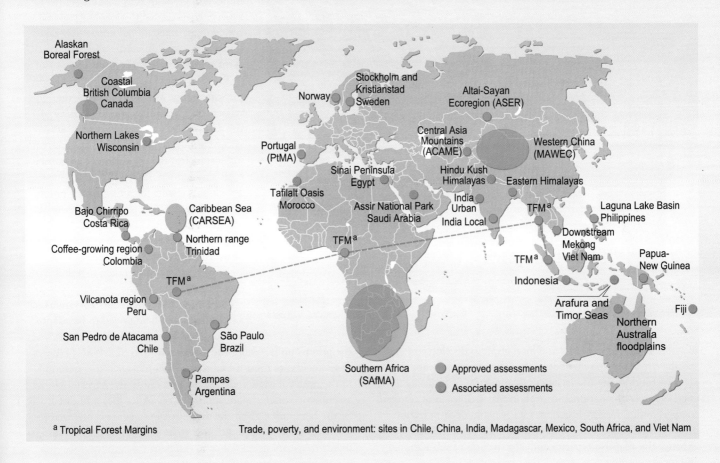

ᵃ Tropical Forest Margins   Trade, poverty, and environment: sites in Chile, China, India, Madagascar, Mexico, South Africa, and Viet Nam

construction by the state followed by colonizing migrant settlers, who in turn practice slash-and-burn agriculture, is most frequent in lowland areas of Latin America, especially in the Amazon Basin. Pasture creation for cattle ranching is causing deforestation almost exclusively in the humid lowland regions of mainland South America. The spontaneous expansion of smallholder agriculture and fuelwood extraction for domestic uses are important causes of deforestation in Africa.

The assessments identified inequities in the distribution of the costs and benefits of ecosystem change, which are often displaced to other places or future generations (SG.SDM). For example, the increase in urbanization in countries like Portugal is generating pressures on ecosystems and services in rural areas. The increase in international trade is also generating additional pressures around the world, illustrated by the cases of the mining industries in Chile and Papua New Guinea. In some situations, the costs of transforming ecosystems are simply deferred to future generations. An example reported widely across sub-global

assessments in different parts of the world is tropical deforestation, which caters to current needs but leads to a reduced capacity to supply services in the future.

Declining ecosystem trends have sometimes been mitigated by innovative local responses. The "threats" observed at an aggregated, global level may be both overestimated and underestimated from a sub-global perspective (SG.SDM). Assessments at an aggregated level often fail to take into account the adaptive capacity of sub-global actors. Through collaboration in social networks, actors can develop new institutions and reorganize to mitigate declining conditions. On the other hand, sub-global actors tend to neglect drivers that are beyond their reach of immediate influence when they craft responses. Hence, it is crucial for decision-makers to develop institutions at the global, regional, and national levels that strengthen the adaptive capacity

are not necessarily seen to be of value locally. Similarly, services of local importance, such as the cultural benefits of ecosystems, the availability of manure for fuel and fertilizer, or the presence of non-wood forest products, are often not seen as important globally. Responses designed to achieve goals related to global or regional concerns are likely to fail unless they take into account the different values and concerns motivating local communities.

There is evidence that including multiple knowledge systems increases the relevance, credibility, and legitimacy of the assessment results for some users (SG.SDM). For example, in Bajo Chirripó in Costa Rica, the involvement of nonscientists added legitimacy and relevance to assessment results for a number of potential users at the local level. In many of the sub-global assessments, however, local resource users were one among many groups of decision-makers, so the question of legitimacy needs to be taken together with that of empowerment.

Integrated assessments of ecosystems and human well-being need to be adapted to the specific needs and characteristics of the groups undertaking the assessment (SG.SDM, SG11.ES). Assessments are most useful to decision-makers if they respond to the needs of those individuals. As a result, the MA sub-global assessments differed significantly in the issues they addressed. At the same time, given the diversity of assessments involved in the MA, the basic approach had to be adapted by different assessments to ensure its relevance to different user groups. (See Box 6.1.) Several community-based assessments adapted the MA framework to allow for more dynamic interplays between variables, to capture fine-grained patterns and processes in complex systems, and to leave room for a more spiritual worldview. In Peru and Costa Rica, for example, other conceptual frameworks were used that incorporated both the MA principles and local cosmologies. In southern Africa, various frameworks were used in parallel to offset the shortcomings of the MA framework for community assessments. These modifications and adaptations of the framework are an important outcome of the MA.

of actors at the sub-national and local levels to develop context-specific responses that do address the full range of relevant drivers. The Biodiversity Management Committees in India are a good example of a national institution that enables local actors to respond to biodiversity loss. This means neither centralization nor decentralization but institutions at multiple levels that enhance the adaptive capacity and effectiveness of sub-national and local responses.

Multiscale assessments offer insights and results that would otherwise be missed (SG.SDM). The variability among sub-global assessments in problem definition, objectives, scale criteria, and systems of explanation increased at finer scales of assessment (for example, social equity issues became more visible from coarser to finer scales of assessment). The role of biodiversity as a risk avoidance mechanism for local communities is frequently hidden until local assessments are conducted (as in the Indian local, Sinai, and Southern African livelihoods studies).

Failure to acknowledge that stakeholders at different scales perceive different values in various ecosystem services can lead to unworkable and inequitable policies or programs at all scales (SGWG). Ecosystem services that are of considerable importance at global scales, such as carbon sequestration or waste regulation,

Box 6.1 Local Adaptations of MA Conceptual Framework (SG.SDM)

The MA framework was applied in a wide range of assessments at multiple scales. Particularly for the more local assessments, the framework needed to be adapted to better reflect the needs and concerns of local communities. In the case of an assessment conducted by and for indigenous communities in the Vilcanota region of Peru, the framework had to be recreated from a base with the Quechua understanding of ecological and social relationships. (See Figure.) Within the Quechua vision of the cosmos, concepts such as reciprocity (Ayni), the inseparability of space and time, and the cyclical nature of all processes (Pachakuti) are important components of the Inca definition of ecosystems. Love (Munay) and working (Llankay) bring humans to a higher state of knowledge (Yachay) about their surroundings and are therefore key concepts linking Quechua communities to the natural world. Ayllu represents the governing institutions that regulate interactions between all living beings.

The resulting framework has similarities with the MA Conceptual Framework, but the divergent features are considered to be important to the Quechua people conducting the assessment. The Vilcanota conceptual framework also includes multiple scales (Kaypacha, Hananpacha, Ukupacha); however, these represent both spatial scales and the cyclical relationship between the past, present, and future. Inherent in this concept of space and time is the adaptive capacity of the Quechua people, who welcome change and have become resilient to it through an adaptive learning process. (It is recognized that current rates of change may prove challenging to the adaptive capacities of the communities.) The cross shape of the Vilcanota framework diagram represents the "Chakana," the most recognized and sacred shape to Quechua people, and orders the

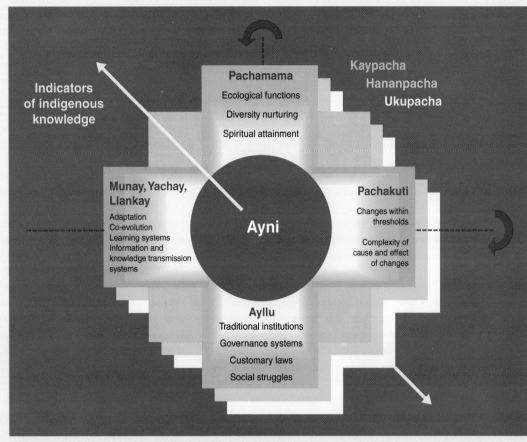

Source: Millennium Ecosystem Assessment, Vilcanota Sub-global Assessment

world through deliberative and collective decision-making that emphasizes reciprocity (Ayni). Pachamama is similar to a combination of the "ecosystem goods and services" and "human well-being" components of the MA framework. Pachakuti is similar to the MA "drivers" (both direct and indirect). Ayllu (and Munay, Yachay, and Llankay) may be seen as responses and are more organically integrated into the cyclic process of change and adaptation.

In the Vilcanota assessment, the Quechua communities directed their work process to assess the conditions and trends of certain aspects of the Pachamama (focusing on water, soil, and agrobiodiversity), how these goods and services are changing, the reasons behind the changes, the effects on the other elements of the Pachamama, how the communities have adapted and are adapting to the changes, and the state of resilience of the Quechua principles and institutions for dealing with these changes in the future.

Developing the local conceptual framework from a base of local concepts and principles, as opposed to simply translating the MA framework into local terms, has allowed local communities to take ownership of their assessment process and given them the power both to assess the local environment and human populations using their own knowledge and principles of well-being and to seek responses to problems within their own cultural and spiritual institutions.

7. *What is known about time scales, inertia, and the risk of nonlinear changes in ecosystems?*

The time scale of change refers to the time required for the effects of a perturbation of a process to be expressed. Time scales relevant to ecosystems and their services are shown in Figure 7.1. Inertia refers to the delay or slowness in the response of a system to factors altering their rate of change, including continuation of change in the system after the cause of that change has been removed. Resilience refers to the amount of disturbance or stress that a system can absorb and still remain capable of returning to its predisturbance state.

Time Scales and Inertia

Many impacts of humans on ecosystems (both harmful and beneficial) are slow to become apparent; this can result in the costs associated with ecosystem changes being deferred to future generations. For example, excessive phosphorus is accumulating in many agricultural soils, threatening rivers, lakes, and coastal oceans with increased eutrophication. Yet it may take years or decades for the full impact of the phosphorus to become apparent through erosion and other processes (S7.3.2). Similarly, the use of groundwater supplies can exceed the recharge rate for some time before costs of extraction begin to grow significantly. In general, people manage ecosystems in a manner that increases short-term benefits; they may not be aware of, or may ignore, costs that are not readily and immediately apparent. This has the inequitable result of increasing current benefits at costs to future generations.

Different categories of ecosystem services tend to change over different time scales, making it difficult for managers to evaluate trade-offs fully. For example, supporting services such as soil formation and primary production and regulating services such as water and disease regulation tend to change over much longer time scales than provisioning services. As a consequence, impacts on more slowly changing supporting and regulating services are often overlooked by managers in pursuit of increased use of provisioning services (S12.ES).

The inertia of various direct and indirect drivers differs considerably, and this strongly influences the time frame for solving ecosystem-related problems once they are identified (RWG, S7). For some drivers, such as the overharvest of particular species, lag times are rather short, and the impact of the driver can be minimized or halted within short time frames. For others, such as nutrient loading and, especially, climate change, lag times are much longer, and the impact of the driver cannot be lessened for years or decades.

Significant inertia exists in the process of species extinctions that result from habitat loss; even if habitat loss were to end today, it would take hundreds of years for species numbers to reach a new and lower equilibrium due to the habitat changes that have taken place in the last centuries (S10). Most species that will go extinct in the next several centuries will be driven to extinction as a result of loss or degradation of their habitat (either through land cover changes or increasingly through climate changes). Habitat loss can lead to rapid extinction of some species (such as those with extremely limited ranges); but for many species, extinction will only occur after many generations, and long-lived species such as some trees could persist for centuries before ultimately going extinct. This "extinction debt" has important implications. First, while reductions in the rate of habitat loss will protect certain species and have significant long-term benefits for species survival in the aggregate, the impact on rates of extinction over the next 10–50 years is likely to be small (*medium certainty*). Second, until a species does go extinct, opportunities exist for it to be recovered to a viable population size.

Nonlinear Changes in Ecosystems

Nonlinear changes, including accelerating, abrupt, and potentially irreversible changes, have been commonly encountered in ecosystems and their services. Most of the time, change in ecosystems and their services is gradual and incremental. Most of these gradual changes are detectable and predictable, at least in principle (*high certainty*) (S.SDM). However, many examples exist of nonlinear and sometimes abrupt changes in ecosystems. In these cases, the ecosystem may change gradually until a particular pressure on it reaches a threshold, at which point changes occur relatively rapidly as the system shifts to a new state. Some of these nonlinear changes can be very large in magnitude and have substantial impacts on human well-being. Capabilities for predicting some nonlinear changes are improving, but for most ecosystems and for most potential nonlinear changes, while science can often warn of increased risks of change, it cannot predict the thresholds where the change will be encountered (C6.2, S13.4). Numerous examples exist of nonlinear and relatively abrupt changes in ecosystems:

■ *Disease emergence* (S13.4): Infectious diseases regularly exhibit nonlinear behavior. If, on average, each infected person infects at least one other person, then an epidemic spreads, while if the infection is transferred on average to less than one person the epidemic dies out. High human population densities in close contact with animal reservoirs of infectious disease facilitate rapid exchange of pathogens, and if the threshold rate of infection is achieved—that is, if each infected person on average transmits the infection to at least one other person—the resulting infectious agents can spread quickly through a worldwide contiguous, highly mobile, human population with few barriers to transmis-

Figure 7.1. Characteristic Time and Space Scales Related to Ecosystems and Their Services

Note: For comparison, this Figure includes references to time and space scales cited in the Synthesis Report of the IPCC Third Assessment Report. (IPCC TAR, C4 Fig 4.15, C4.4.2, CF7, S7)

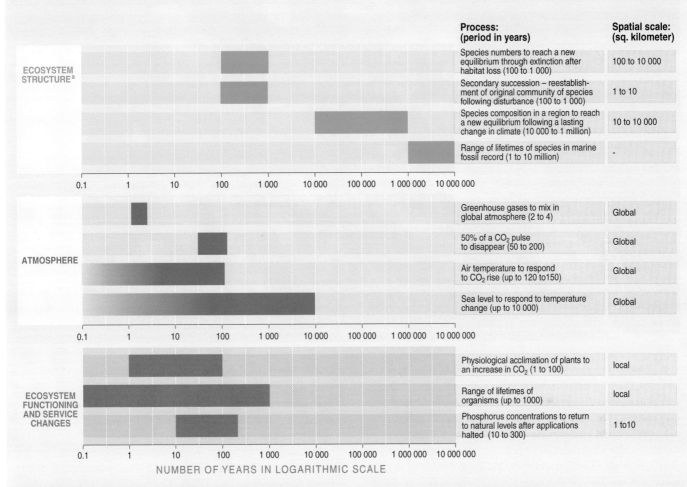

Process: (period in years)	Spatial scale: (sq. kilometer)
ECOSYSTEM STRUCTURE[a]	
Species numbers to reach a new equilibrium through extinction after habitat loss (100 to 1 000)	100 to 10 000
Secondary succession – reestablishment of original community of species following disturbance (100 to 1 000)	1 to 10
Species composition in a region to reach a new equilibrium following a lasting change in climate (10 000 to 1 million)	10 to 10 000
Range of lifetimes of species in marine fossil record (1 to 10 million)	-
ATMOSPHERE	
Greenhouse gases to mix in global atmosphere (2 to 4)	Global
50% of a CO_2 pulse to disappear (50 to 200)	Global
Air temperature to respond to CO_2 rise (up to 120 to150)	Global
Sea level to respond to temperature change (up to 10 000)	Global
ECOSYSTEM FUNCTIONING AND SERVICE CHANGES	
Physiological acclimation of plants to an increase in CO_2 (1 to 100)	local
Range of lifetimes of organisms (up to 1000)	local
Phosphorus concentrations to return to natural levels after applications halted (10 to 300)	1 to10

NUMBER OF YEARS IN LOGARITHMIC SCALE

[a] The ecosystem structure category includes also the "range size of vertabrate species" for which the time scale is not available. The spatial scale goes from 0.1 to 100 million square kilometers.

Sources: IPCC, Millennium Ecosystem Assessment

sion. The almost instantaneous outbreak of SARS in different parts of the world is an example of such potential, although rapid and effective action contained its spread. During the 1997/98 El Niño, excessive flooding caused cholera epidemics in Djibouti, Somalia, Kenya, Tanzania, and Mozambique. Warming of the African Great Lakes due to climate change may create conditions that increase the risk of cholera transmission in surrounding countries (C14.2.1). An event similar to the 1918 Spanish flu pandemic, which is thought to have killed 20–40 million people worldwide, could now result in over 100 million deaths within a single year. Such a catastrophic event, the possibility of which is being seriously considered by the epidemiological community, would probably lead to severe economic disruption and possibly even rapid collapse in a world economy dependent on fast global exchange of goods and services.

■ *Algal blooms and fish kills* (S13.4): Excessive nutrient loading fertilizes freshwater and coastal ecosystems. While small increases in nutrient loading often cause little change in many ecosystems, once a threshold of nutrient loading is achieved, the changes can be abrupt and extensive, creating harmful algal blooms (including blooms of toxic species) and often leading to the domination of the ecosystem by one or a few species. Severe nutrient overloading can lead to the formation of oxygen-depleted zones, killing all animal life.

■ *Fisheries collapses* (C18): Fish population collapses have been commonly encountered in both freshwater and marine fisheries. Fish populations are generally able to withstand some level of catch with a relatively small impact on their overall population size. As the catch increases, however, a threshold is reached after which too few adults remain to produce enough offspring to support that level of harvest, and the population may drop abruptly to a much smaller size. For example, the Atlantic cod stocks of the east coast of Newfoundland collapsed in 1992, forcing the closure of the fishery after hundreds of years of exploitation, as shown in Figure 3.4 (CF2 Box 2.4). Most important, the stocks may take years to recover or not recover at all, even if harvesting is significantly reduced or eliminated entirely.

■ *Species introductions and losses:* Introductions (or removal) of species can cause nonlinear changes in ecosystems and their services. For example, the introduction of the zebra mussel (see photo above) into U.S. aquatic systems resulted in the extirpation of native clams in Lake St. Clair, large changes in energy flow and ecosystem function, and annual costs of $100 million to the power industry and other users (S12.4.8). The introduction of the comb jelly fish (*Mnemiopsis leidyi*) in the Black Sea caused the loss of 26 major fisheries species and has been implicated (along with other factors) in subsequent growth of the anoxic "dead zone" (C28.5). The loss of the sea otters from many coastal ecosystems on the Pacific Coast of North America due to hunting led to the booming populations of sea urchins (a prey species for otters) which in turn led to the loss of kelp forests (which are eaten by urchins).

■ *Changes in dominant species in coral ecosystems:* Some coral reef ecosystems have undergone sudden shifts from coral-dominated to algae-dominated reefs. The trigger for such phase shifts, which are essentially irreversible, is usually multifaceted and includes increased nutrient input leading to eutrophic conditions, and removal of herbivorous fishes that maintain the balance between corals and algae. Once a threshold is reached, the change in the ecosystem takes place within months and the resulting ecosystem, although stable, is less productive and less diverse. One well-studied example is the sudden switch in 1983 from coral to algal domination of Jamaican reef systems. This followed several centuries of overfishing of herbivores, which left the control of algal cover almost entirely dependent on a single species of sea urchin, whose populations collapsed when exposed to a species-specific pathogen. As a result, Jamaica's reefs shifted (apparently irreversibly) to a new low-diversity, algae-dominated state with very limited capacity to support fisheries (C4.6).

■ *Regional climate change* (C13.3): The vegetation in a region influences climate through albedo (reflectance of radiation from the surface), transpiration (flux of water from the ground to the atmosphere through plants), and the aerodynamic properties of

the surface. In the Sahel region of North Africa, vegetation cover is almost completely controlled by rainfall. When vegetation is present, rainfall is quickly recycled, generally increasing precipitation and, in turn, leading to a denser vegetation canopy. Model results suggest that land degradation leads to a substantial reduction in water recycling and may have contributed to the observed trend in rainfall reduction in the region over the last 30 years. In tropical regions, deforestation generally leads to decreased rainfall. Since forest existence crucially depends on rainfall, the relationship between tropical forests and precipitation forms a positive feedback that, under certain conditions, theoretically leads to the existence of two steady states: rainforest and savanna (although some models suggest only one stable climate-vegetation state in the Amazon).

There is *established but incomplete* evidence that changes being made in ecosystems are increasing the likelihood of nonlinear and potentially high-impact, abrupt changes in physical and biological systems that have important consequences for human well-being (C6, S3, S13.4, S.SDM). The increased likelihood of these events stems from the following factors:

■ *On balance, changes humans are making to ecosystems are reducing the resilience of the ecological components of the systems* (*established but incomplete*) (C6, S3, S12). Genetic and species diversity, as well as spatial patterns of landscapes, environmental fluctuations, and temporal cycles with which species evolved, generate the resilience of ecosystems. Functional groups of species contribute to ecosystem processes and services in similar ways. Diversity among functional groups increases the flux of ecosystem processes and services (*established but incomplete*). Within functional groups, species respond differently to environmental fluctuations. This response diversity derives from variation in the response of species to environmental drivers, heterogeneity in species distributions, differences in ways that species use seasonal cycles or disturbance patterns, or other mechanisms. Response diversity enables ecosystems to adjust in changing environments, altering biotic structure in ways that maintain processes and services (*high certainty*) (S.SDM). The loss of biodiversity that is now taking place thus tends to reduce the resilience of ecosystems.

■ *There are growing pressures from various drivers* (S7, SG7.5). Threshold changes in ecosystems are not uncommon, but they are infrequently encountered in the absence of human-caused pressures on ecosystems. Many of these pressures are now growing. Increased fish harvests raise the likelihood of fisheries collapses; higher rates of climate change boost the potential for species extinctions; increased introductions of nitrogen and phosphorus into the environment make the eutrophication of aquatic ecosystems more likely; as human populations become more mobile, more and more species are being introduced into new habitats, and this increases the chance of harmful pests emerging in those regions.

The growing bushmeat trade poses particularly significant threats associated with nonlinear changes, in this case accelerating rates of change (C8.3, S.SDM, C14). Growth in the use and trade of bushmeat is placing increasing pressure on many species, particularly in Africa and Asia. While population size of harvested species may decline gradually with increasing harvest for some time, once the harvest exceeds sustainable levels, the rate of decline of populations of the harvested species will tend to accelerate. This could place them at risk of extinction and also reduce the food supply of the people dependent on these resources. Finally, the bushmeat trade involves relatively high levels of interaction between humans and some relatively closely related wild animals that are eaten. Again, this increases the risk of a nonlinear change, in this case the emergence of new and serious pathogens. Given the speed and magnitude of international travel today, new pathogens could spread rapidly around the world.

A potential nonlinear response, currently the subject of intensive scientific research, is the atmospheric capacity to cleanse itself of air pollution (in particular, hydrocarbons and reactive nitrogen compounds) (C.SDM). This capacity depends on chemical reactions involving the hydroxyl radical, the atmospheric concentration of which has declined by about 10% (*medium certainty*) since preindustrial times.

Once an ecosystem has undergone a nonlinear change, recovery to the original state may take decades or centuries and may sometimes be impossible. For example, the recovery of overexploited fisheries that have been closed to fishing is quite variable. Although the cod fishery in Newfoundland has been closed for 13 years (except for a small inshore fishery between 1998 and 2003), there have been few signs of a recovery, and many scientists are not optimistic about its return in the foreseeable future (C18.2.6). On the other hand, the North Sea Herring fishery collapsed due to overharvesting in the late 1970s, but it recovered after being closed for four years (C18).

8. *What options exist to manage ecosystems sustainably?*

It is a major challenge to reverse the degradation of ecosystems while meeting increasing demands for their services. But this challenge can be met. Three of the four MA scenarios show that changes in policies, institutions, and practices can mitigate some of the negative consequences of growing pressures on ecosystems, although the changes required are large and not currently under way (S.SDM). As noted in Key Question 5, in three of the four MA scenarios at least one of the three categories of provisioning, regulating, and cultural services is in better condition in 2050 than in 2000, although biodiversity loss continues at high rates in all scenarios. The scale of interventions that results in these positive outcomes, however, is very significant. The interventions include major investments in environmentally sound technology, active adaptive management, proactive actions to address environmental problems before their full consequences are experienced, major investments in public goods (such as education and health), strong action to reduce socioeconomic disparities and eliminate poverty, and expanded capacity of people to manage ecosystems adaptively.

More specifically, in *Global Orchestration* trade barriers are eliminated, distorting subsidies are removed, and a major emphasis is placed on eliminating poverty and hunger. In *Adapting Mosaic,* by 2010 most countries are spending close to 13% of their GDP on education (compared with an average of 3.5% in 2000), and institutional arrangements to promote transfer of skills and knowledge among regional groups proliferate. In *TechnoGarden,* policies are put in place to provide payment to individuals and companies that provide or maintain the provision of ecosystem services. For example, in this scenario, by 2015 roughly 50% of European agriculture and 10% of North American agriculture is aimed at balancing the production of food with the production of other ecosystem services. Under this scenario, significant advances occur in the development of environmental technologies to increase production of services, create substitutes, and reduce harmful trade-offs.

Past actions to slow or reverse the degradation of ecosystems have yielded significant benefits, but these improvements have generally not kept pace with growing pressures and demands. Although most ecosystem services assessed in the MA are being degraded, the extent of that degradation would have been much greater without responses implemented in past decades. For example, more than 100,000 protected areas (including strictly protected areas such as national parks as well as areas managed for the sustainable use of natural ecosystems, including timber harvest or wildlife harvest) covering about 11.7% of the terrestrial surface have now been established (R5.2.1). These play an important role in the conservation of biodiversity and ecosystem services, although important gaps in the distribution of protected areas remain, particularly in marine and freshwater systems.

Technological advances have also helped to lessen the rate of growth in pressure on ecosystems caused per unit increase in demand for ecosystem services. For all developing countries, for instance, yields of wheat, rice, and maize rose between 109% and 208% in the past 40 years. Without this increase, far more habitat would have been converted to agriculture during this time.

An effective set of responses to ensure the sustainable management of ecosystems must address the drivers presented in Key Question 4 and overcome barriers related to (RWG):

- inappropriate institutional and governance arrangements, including the presence of corruption and weak systems of regulation and accountability;
- market failures and the misalignment of economic incentives;
- social and behavioral factors, including the lack of political and economic power of some groups (such as poor people, women, and indigenous groups) who are particularly dependent on ecosystem services or harmed by their degradation;
- underinvestment in the development and diffusion of technologies that could increase the efficiency of use of ecosystem services and reduce the harmful impacts of various drivers of ecosystem change; and
- insufficient knowledge (as well as the poor use of existing knowledge) concerning ecosystem services and management, policy, technological, behavioral and institutional responses that could enhance benefits from these services while conserving resources.

All these barriers are compounded by weak human and institutional capacity related to the assessment and management of ecosystem services, underinvestment in the regulation and management of their use, lack of public awareness, and lack of awareness among decision-makers of the threats posed by the degradation of ecosystem services and the opportunities that more sustainable management of ecosystems could provide.

The MA assessed 74 response options for ecosystem services, integrated ecosystem management, conservation and sustainable use of biodiversity, and climate change. (See Appendix B.) Many of these options hold significant promise for conserving or sustainably enhancing the supply of ecosystem services. Examples of promising responses that address the barriers just described are presented in the remainder of this section (RWG, R2). The stakeholder groups that would need to take decisions to implement each response are indicated as follows: G for government, B for business and industry, and N for nongovernmental organizations and other civil society organizations such as community-based and indigenous peoples organizations.

Institutions and Governance

Changes in institutional and environmental governance frameworks are sometimes required in order to create the enabling conditions for effective management of ecosystems, while in other cases existing institutions could meet these needs but face significant barriers. Many existing institutions at both the global and the national level have the mandate to address the degradation of ecosystem services but face a variety of challenges in doing so related to the need for greater cooperation across sectors and the need for coordinated responses at multiple scales. However, since a number of the issues identified in this assessment are recent concerns and were not specifically taken into account in the design of today's institutions, changes in existing institutions and the development of new ones may sometimes be needed, particularly at the national scale.

In particular, existing national and global institutions are not well designed to deal with the management of open access resources, a characteristic of many ecosystem services. Issues of ownership and access to resources, rights to participation in decision-making, and regulation of particular types of resource use or discharge of wastes can strongly influence the sustainability of ecosystem management and are fundamental determinants of who wins and who loses from changes in ecosystems. Corruption—a major obstacle to effective management of ecosystems—also stems from weak systems of regulation and accountability.

Promising interventions include:

■ *Integration of ecosystem management goals within other sectors and within broader development planning frameworks* (G). The most important public policy decisions affecting ecosystems are often made by agencies and in policy arenas other than those charged with protecting ecosystems. Ecosystem management goals are more likely to be achieved if they are reflected in decisions in other sectors and in national development strategies. For example, the Poverty Reduction Strategies prepared by developing-country governments for the World Bank and other institutions strongly shape national development priorities, but in general these have not taken into account the importance of ecosystems to improving the basic human capabilities of the poorest (R17.ES).

■ *Increased coordination among multilateral environmental agreements and between environmental agreements and other international economic and social institutions* (G). International agreements are indispensable for addressing ecosystem-related concerns that span national boundaries, but numerous obstacles weaken their current effectiveness (R17.2). The limited, focused nature of the goals and mechanisms included in most bilateral and multilateral environmental treaties does not address the broader issue of ecosystem services and human well-being. Steps are now being taken to increase coordination among these treaties, and this could help broaden the focus of the array of instruments. However, coordination is also needed between the multilateral environmental agreements and the more politically powerful international legal institutions, such as economic and trade agreements, to ensure that they are not acting at cross-purposes (R.SDM). And implementation of these agreements also needs to be coordinated among relevant institutions and sectors at the national level.

■ *Increased transparency and accountability of government and private-sector performance in decisions that affect ecosystems, including through greater involvement of concerned stakeholders in decision-making* (G, B, N) (RWG, SG9). Laws, policies, institutions, and

markets that have been shaped through public participation in decision-making are more likely to be effective and perceived as just. For example, degradation of freshwater and other ecosystem services generally have a disproportionate impact on those who are, in various ways, excluded from participation in the decision-making process (R7.2.3). Stakeholder participation also contributes to the decision-making process because it allows a better understanding of impacts and vulnerability, the distribution of costs and benefits associated with trade-offs, and the identification of a broader range of response options that are available in a specific context. And stakeholder involvement and transparency of decision-making can increase accountability and reduce corruption.

■ *Development of institutions that devolve (or centralize) decision-making to meet management needs while ensuring effective coordination across scales* (G, B, N) (RWG). Problems of ecosystem management have been exacerbated by both overly centralized and overly decentralized decision-making. For example, highly centralized forest management has proved ineffective in many countries, and efforts are now being made to move responsibility to lower levels of decision-making either within the natural resources sector or as part of broader decentralization of governmental responsibilities. At the same time, one of the most intractable problems of ecosystem management has been the lack of alignment between political boundaries and units appropriate for the management of ecosystem goods and services. Downstream communities may not have access to the institutions through which upstream actions can be influenced; alternatively, downstream communities or countries may be stronger politically than upstream regions and may dominate control of upstream areas without addressing upstream needs. A number of countries, however, are now strengthening regional institutions for the management of transboundary ecosystems (such as the Danube River, the Mekong River Commission, East African cooperation on Lake Victoria, and the Amazon Cooperation Treaty Organization).

■ *Development of institutions to regulate interactions between markets and ecosystems* (G) (RWG). The potential of policy and market reforms to improve ecosystem management are often constrained by weak or absent institutions. For example, the potential of the Clean Development Mechanism established

under the Framework Convention on Climate Change to provide financial support to developing countries in return for greenhouse gas reductions, which would realize climate and biodiversity benefits through payments for carbon sequestration in forests, is constrained by unclear property rights, concerns over the permanence of reductions, and lack of mechanisms for resolving conflicts. Moreover, existing regulatory institutions often do not have ecosystem protection as a clear mandate. For example, independent regulators of privatized water systems and power systems do not necessarily promote resource use efficiency and renewable supply. There is a continuing importance of the role of the state to set and enforce rules even in the context of privatization and market-led growth.

■ *Development of institutional frameworks that promote a shift from highly sectoral resource management approaches to more integrated approaches* (G, B) (R15.ES, R12.ES, R11.ES). In most countries, separate ministries are in charge of different aspects of ecosystems (such as ministries of environment, agriculture, water, and forests) and different drivers of change (such as ministries of energy, transportation, development, and trade). Each of these ministries has control over different aspects of ecosystem management. As a result, there is seldom the political will to develop effective ecosystem management strategies, and competition among the ministries can often result in policy choices that are detrimental to ecosystems. Integrated responses intentionally and actively address ecosystem services and human well-being simultaneously, such as integrated coastal zone management, integrated river basin management, and national sustainable development strategies. Although the potential for integrated

RON GILING/PETER ARNOLD, INC

responses is high, numerous barriers have limited their effectiveness: they are resource-intensive, but the potential benefits can exceed the costs; they require multiple instruments for their implementation; and they require new institutional and governance structures, skills, knowledge, and capacity. Thus far, the results of implementation of integrated responses have been mixed in terms of ecological, social, and economic impacts.

Economics and Incentives

Economic and financial interventions provide powerful instruments to regulate the use of ecosystem goods and services (C5 Box 5.2). Because many ecosystem services are not traded in markets, markets fail to provide appropriate signals that might otherwise contribute to the efficient allocation and sustainable use of the services. Even if people are aware of the services provided by an ecosystem, they are neither compensated for providing these services nor penalized for reducing them. In addition, the people harmed by the degradation of ecosystem services are often not the ones who benefit from the actions leading to their degradation, and so those costs are not factored into management decisions. A wide range of opportunities exists to influence human behavior to address this challenge in the form of economic and financial instruments. Some of them establish markets; others work through the monetary and financial interests of the targeted social actors; still others affect relative prices.

Market mechanisms can only work if supporting institutions are in place, and thus there is a need to build institutional capacity to enable more widespread use of these mechanisms (R17). The adoption of economic instruments usually requires a legal framework, and in many cases the choice of a viable and effective economic intervention mechanism is determined by the socioeconomic context. For example, resource taxes can be a powerful instrument to guard against the overexploitation of an ecosystem service, but an effective tax scheme requires well-established and reliable monitoring and tax collection systems. Similarly, subsidies can be effective to introduce and implement certain technologies or management procedures, but they are inappropriate in settings that lack the transparency and accountability needed to prevent corruption. The establishment of market mechanisms also often involves explicit decisions about wealth distribution and resource allocation, when, for example, decisions are made to establish private property rights for resources that were formerly considered common pool resources. For that reason, the inappropriate use of market mechanisms can further exacerbate problems of poverty.

Promising interventions include:

■ *Elimination of subsidies that promote excessive use of ecosystem services* (*and, where possible, transfer of these subsidies to payments for nonmarketed ecosystem services*) (G) (S7.ES). Subsidies paid to the agricultural sectors of OECD countries between 2001 and 2003 averaged over $324 billion annually, or one third the global value of agricultural products in 2000. Many countries outside the OECD also have inappropriate subsidies. A significant proportion of this total involves production subsidies that lead to greater food production in countries with subsidies than the global market conditions warrant, that promote the overuse of water, fertilizers, and pesticides, and that reduce the profitability of agriculture in developing countries. They also increase land values, adding to landowners' resistance to subsidy reductions. On the social side, agricultural subsidies make farmers overly dependent on taxpayers for their livelihood, change wealth distribution and social composition by benefiting large corporate farms to the detriment of smaller family farms, and contribute to the dependence of large segments of the developing world on aid. Finally, it is not clear that these policies achieve one of their primary targets—supporting farmers' income. Only about a quarter of the total expenses in price supports translate into additional income for farm households.

Similar problems are created by fishery subsidies, which for the OECD countries were estimated at $6.2 billion in 2002, or about 20% of the gross value of production that year (C8.4.1). Subsidies on fisheries, apart from their distributional impacts, affect the management of resources and their sustainable use by encouraging overexploitation of the resource, thereby worsening the common property problem present in fisheries. Although some indirect subsidies, such as payments for the withdrawal of individual transferable harvest quotas, could have a positive impact on fisheries management, the majority of subsidies have a negative effect. Inappropriate subsidies are also common in sectors such as water and forestry.

Although removal of production subsidies would produce net benefits, it would not occur without costs. The farmers and fishers benefiting directly from the subsidies would suffer the most immediate losses, but there would also be indirect effects on ecosystems both locally and globally. In some cases it may be possible to transfer production subsides to other activities that promote ecosystem stewardship, such as payment for the provision or enhancement of regulatory or supporting services. Compensatory mechanisms may be needed for the poor who are adversely affected by the immediate removal of subsidies (R17.5). Reduced subsidies within the OECD may lessen pressures on some ecosystems in those countries, but they could lead to more rapid conversion and intensification of land for agriculture in developing countries and would thus need to be accompanied by policies to minimize the adverse impacts on ecosystems there.

■ *Greater use of economic instruments and market-based approaches in the management of ecosystem services* (G, B, N) (RWG). Economic instruments and market mechanisms with the potential to enhance the management of ecosystem services include:

- *Taxes or user fees for activities with "external" costs* (trade-offs not accounted for in the market). These instruments create an incentive that lessens the external costs and provides revenues that can help protect the damaged ecosystem services. Examples include taxes on excessive application of nutrients or ecotourism user fees.

- *Creation of markets, including through cap-and-trade systems.* Ecosystem services that have been treated as "free" resources, as is often the case for water, tend to be used wastefully. The establishment of markets for the services can both increase the incentives for their conservation and increase the economic efficiency of their allocation if supporting legal and economic institutions are in place. However, as noted earlier, while markets will increase the efficiency of the use of the resource, they can have harmful effects on particular groups of users who may inequitably affected by the change (R17). The combination of regulated emission caps, coupled with market mechanisms for trading pollution rights, often provides an efficient means of reducing emissions harmful to ecosystems. For example, nutrient trading systems may be a low-cost way to reduce water pollution in the United States (R7 Box 7.3).

One of the most rapidly growing markets related to ecosystem services is the carbon market. (See Figure 8.1.) Approximately 64 million tons of carbon dioxide equivalent were exchanged through projects from January to May 2004, nearly as much as during all of 2003 (78 million tons) (C5 Box 5.2). The value of carbon dioxide trades in 2003 was approximately $300 million. About one quarter of the trades (by volume of CO_2 equivalents) involve investment in ecosystem services (hydropower or biomass). The World Bank has established a fund with a capital of $33.3 million (as of January 2005) to invest in afforestation and reforestation projects that sequester or conserve carbon in forest and

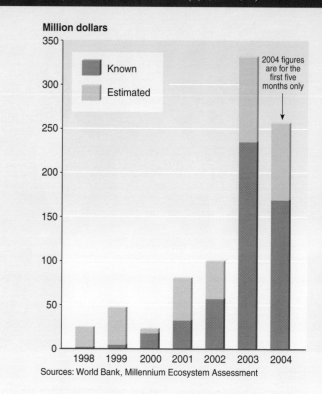

Figure 8.1. TOTAL CARBON MARKET VALUE PER YEAR (in million dollars nominal) (C5 Box 5.1)

Sources: World Bank, Millennium Ecosystem Assessment

agroecosystems while promoting biodiversity conservation and poverty alleviation. It is *speculated* that the value of the global carbon emissions trading markets may reach $10 billion to $44 billion in 2010 (and involve trades totaling 4.5 billion tons of carbon dioxide or equivalent).

- *Payment for ecosystem services.* Mechanisms can be established to enable individuals, firms, or the public sector to pay resource owners to provide particular services. For example, in New South Wales, Australia, associations of farmers purchase salinity credits from the State Forests Agency, which in turn contracts with upstream landholders to plant trees, which reduce water tables and store carbon. Similarly, in 1996 Costa Rica established a nationwide system of conservation payments to induce landowners to provide ecosystem services. Under this program, the government brokers contracts between international and domestic "buyers" and local "sellers" of sequestered carbon, biodiversity, watershed services, and scenic beauty. By 2001, more than 280,000 hectares of forests had been incorporated into the program at a cost of about $30 million, with pending applications covering an additional 800,000 hectares (C5 Box 5.2).

Other innovative conservation financing mechanisms include "biodiversity offsets" (whereby developers pay for conservation activities as compensation for unavoidable harm that a project causes to biodiversity). An online news site, the Ecosystem Marketplace, has now been established

by a consortium of institutions to provide information on the development of markets for ecosystem services and the payments for them.

- *Mechanisms to enable consumer preferences to be expressed through markets.* Consumer pressure may provide an alternative way to influence producers to adopt more sustainable production practices in the absence of effective government regulation. For example, certification schemes that exist for sustainable fisheries and forest practices provide people with the opportunity to promote sustainability through their consumer choices. Within the forest sector, forest certification has become widespread in many countries and forest conditions; thus far, however, most certified forests are in temperate regions, managed by large companies that export to northern retailers (R8).

Social and Behavioral Responses

Social and behavioral responses—including population policy; public education; empowerment of communities, women, and youth; and civil society actions—can be instrumental in responding to ecosystem degradation. These are generally interventions that stakeholders initiate and execute through exercising their procedural or democratic rights in efforts to improve ecosystems and human well-being.

Promising interventions include:

- *Measures to reduce aggregate consumption of unsustainably managed ecosystem services* (G, B, N) (RWG). The choices about what individuals consume and how much they consume are influenced not just by considerations of price but also by behavioral factors related to culture, ethics, and values. Behavioral changes that could reduce demand for degraded ecosystem services can be encouraged through actions by governments (such as education and public awareness programs or the promotion of demand-side management), industry (such as improved product labeling or commitments to use raw materials from sources certified as sustainable), and civil society (such as public awareness campaigns). Efforts to reduce aggregate consumption, however, must sometimes incorporate measures to increase the access to and consumption of those same ecosystem services by specific groups such as poor people.

- *Communication and education* (G, B, N) (RWG, R5). Improved communication and education are essential to achieve the objectives of the environmental conventions, the Johannesburg Plan of Implementation, and the sustainable management of natural resources more generally. Both the public and decision-makers can benefit from education concerning ecosystems and human well-being, but education more generally provides tremendous social benefits that can help address many drivers of ecosystem degradation. Barriers to the effective use of communication and education include a failure to use research and apply modern theories of learning and change. While the importance of communication and education is well recognized, providing the human and financial resources to undertake effective work is a continuing barrier.

- *Empowerment of groups particularly dependent on ecosystem services or affected by their degradation, including women, indigenous people, and young people* (G, B, N) (RWG). Despite women's knowledge about the environment and the potential they possess, their participation in decision-making has often been restricted by social and cultural structures. Young people are key stakeholders in that they will experience the longer-term consequences of decisions made today concerning ecosystem services. Indigenous control of traditional homelands can sometimes have environmental benefits, although the primary justification continues to be based on human and cultural rights.

Technological Responses

Given the growing demands for ecosystem services and other increased pressures on ecosystems, the development and diffusion of technologies designed to increase the efficiency of resource use or reduce the impacts of drivers such as climate change and nutrient loading are essential. Technological change has been essential for meeting growing demands for some ecosystem services, and technology holds considerable promise to help meet future growth in demand. Technologies already exist for reducing nutrient pollution at reasonable costs—including technologies to reduce point source emissions, changes in crop management practices, and precision farming techniques to help control the application of fertilizers to a field, for example—but new policies are needed for these tools to be applied on a sufficient scale to slow and ultimately reverse the increase in nutrient loading (recognizing that this global goal must be achieved even while increasing nutrient applications in some regions such as sub-Saharan Africa). Many negative impacts on ecosystems and human well-being have resulted from these technological changes, however (R17.ES). The cost of "retrofitting" technologies once their negative consequences become apparent can be extremely high, so careful assessment is needed prior to the introduction of new technologies.

Promising interventions include:

- *Promotion of technologies that increase crop yields without any harmful impacts related to water, nutrient, and pesticide use* (G, B, N) (R6). Agricultural expansion will continue to be one of the major drivers of biodiversity loss well into the twenty-first century. Development, assessment, and diffusion of technologies that could increase the production of food per unit area sustainably without harmful trade-offs related to excessive use of water, nutrients, or pesticides would significantly lessen pressure on other ecosystem services. Without the intensification that has taken place since 1950, a further 20 million square kilometers of land would have had to be brought into production to achieve today's crop production (C.SDM). The challenge for the future is to similarly reduce the pressure for expansion of agriculture without simultaneously increasing pressures on ecosystem services due to water use, excessive nutrient loading, and pesticide use.

■ *Restoration of ecosystem services* (G, B, N) (RWG, R7.4). Ecosystem restoration activities are now common in many countries and include actions to restore almost all types of ecosystems, including wetlands, forests, grasslands, estuaries, coral reefs, and mangroves. Ecosystems with some features of the ones that were present before conversion can often be established and can provide some of the original ecosystem services (such as pollution filtration in wetlands or timber production from forests). The restored systems seldom fully replace the original systems, but they still help meet needs for particular services. Yet the cost of restoration is generally extremely high in relation to the cost of preventing the degradation of the ecosystem. Not all services can be restored, and those that are heavily degraded may require considerable time for restoration.

■ *Promotion of technologies to increase energy efficiency and reduce greenhouse gas emissions* (G, B) (R13). Significant reductions in net greenhouse gas emissions are technically feasible due to an extensive array of technologies in the energy supply, energy demand, and waste management sectors. Reducing projected emissions will require a portfolio of energy production technologies ranging from fuel switching (coal/oil to gas) and increased power plant efficiency to increased use of renewable energy technologies, complemented by more efficient use of energy in the transportation, buildings, and industry sectors. It will also involve the development and implementation of supporting institutions and policies to overcome barriers to the diffusion of these technologies into the marketplace, increased public and private-sector funding for research and development, and effective technology transfer.

Knowledge and Cognitive Responses

Effective management of ecosystems is constrained both by a lack of knowledge and information concerning different aspects of ecosystems and by the failure to use adequately the information that does exist in support of management decisions. Although sufficient information exists to take many actions that could help conserve ecosystems and enhance human well-being, major information gaps exist. In most regions, for example, relatively little is known about the status and economic value of most ecosystem services, and their depletion is rarely tracked in national economic accounts. Limited information exists about the likelihood of nonlinear changes in ecosystems or the location of thresholds where such changes may be encountered. Basic global data on the extent and trend in different types of ecosystems and land use are surprisingly scarce. Models used to project future environmental and economic conditions have limited capability of incorporating ecological "feedbacks" including nonlinear changes in ecosystems.

At the same time, decision-makers do not use all of the relevant information that is available. This is due in part to institutional failures that prevent existing policy-relevant scientific information from being made available to decision-makers. But it is also due to the failure to incorporate other forms of knowledge and information, such as traditional knowledge and practitioners' knowledge, that are often of considerable value for ecosystem management.

Promising interventions include:

■ *Incorporate both the market and nonmarket values of ecosystems in resource management and investment decisions* (G, B) (RWG). Most resource management and investment decisions are strongly influenced by considerations of the monetary costs and benefits of alternative policy choices. In the case of ecosystem management, however, this often leads to outcomes that are not in the interest of society, since the nonmarketed values of ecosystems may exceed the marketed values. As a result, many existing resource management policies favor sectors such as agriculture, forestry, and fisheries at the expense of the use of these same ecosystems for water supply, recreation, and cultural services that may be of greater economic value. Decisions can be improved if they include the total economic value of alternative management options and involve deliberative mechanisms that bring to bear noneconomic considerations as well.

■ *Use of all relevant forms of knowledge and information in assessments and decision-making, including traditional and practitioners' knowledge* (G, B, N) (RWG, C17.ES). Effective management of ecosystems typically requires "place-based" knowledge—information about the specific characteristics and history of an ecosystem. Formal scientific information is often one source of such information, but traditional knowledge or practitioners' knowledge held by local resource managers can be of equal or greater value. While that knowledge is used in the decisions taken by those who have it, it is too rarely incorporated into other decision-making processes and is often inappropriately dismissed.

■ *Enhance and sustain human and institutional capacity for assessing the consequences of ecosystem change for human well-being and acting on such assessments* (G, B, N) (RWG). Greater technical capacity is needed for agriculture, forest, and fisheries management. But the capacity that exists for these sectors, as limited as it is in many countries, is still vastly greater than the capacity for effective management of other ecosystem services. Because awareness of the importance of these other services has only recently grown, there is limited experience with assessing ecosystem services fully. Serious limits exist in all countries, but especially in developing countries, in terms of the expertise needed in such areas as monitoring changes in ecosystem services, economic valuation or health assessment of ecosystem changes, and policy analysis related to ecosystem services. Even when such assessment information is available, however, the traditional highly sectoral nature of decision-making and resource management makes the implementation of recommendations difficult. This constraint can also be overcome through increased training of individuals in existing institutions and through institutional reforms to build capacity for more integrated responses.

Design of Effective Decision-making Processes

Decisions affecting ecosystems and their services can be improved by changing the processes used to reach those decisions. The context of decision-making about ecosystems is changing rapidly. The new challenge to decision-making is to make effective use of information and tools in this changing context in order to improve the decisions. At the same time, some old challenges must still be addressed. The decision-making process and the actors involved influence the intervention chosen. Decision-making processes vary across jurisdictions, institutions, and cultures. Yet the MA has identified the following elements of decision-making processes related to ecosystems and their services that tend to improve the decisions reached and their outcomes for ecosystems and human well-being (R18.ES):

- Use the best available information, including considerations of the value of both marketed and nonmarketed ecosystem services.
- Ensure transparency and the effective and informed participation of important stakeholders.
- Recognize that not all values at stake can be quantified, and thus quantification can provide a false objectivity in decision processes that have significant subjective elements.
- Strive for efficiency, but not at the expense of effectiveness.
- Consider equity and vulnerability in terms of the distribution of costs and benefits.
- Ensure accountability and provide for regular monitoring and evaluation.
- Consider cumulative and cross-scale effects and, in particular, assess trade-offs across different ecosystem services.

A wide range of deliberative tools (which facilitate transparency and stakeholder participation), information-gathering tools (which are primarily focused on collecting data and opinions), and planning tools (which are typically used to evaluate potential policy options) can assist decision-making concerning ecosystems and their services (R3 Tables 3.6 to 3.8). Deliberative tools include neighborhood forums, citizens' juries, community issues groups, consensus conferences, electronic democracy, focus groups, issue forums, and ecosystem service user forums. Examples of information-gathering tools include citizens' research panels, deliberative opinion polls, environmental impact assessments, participatory rural appraisal, and rapid rural appraisal. Some common planning tools are consensus participation, cost-benefit analysis, multicriteria analysis, participatory learning and action, stakeholder decision analysis, trade-off analysis, and visioning exercises. The use of decision-making methods that adopt a pluralistic perspective is particularly pertinent, since these techniques do not give undue weight to any particular viewpoint. These tools can be used at a variety of scales, including global, sub-global, and local.

A variety of frameworks and methods can be used to make better decisions in the face of uncertainties in data, prediction, context, and scale (R4.5). Commonly used methods include cost-benefit or multicriteria analyses, risk assessment, the precautionary principle, and vulnerability analysis. (See Table 8.1.) All these methods have been able to support optimization exercises,

---

Table 8.1. Applicability of Decision Support Methods and Frameworks (R4 Table 4.1)

Method	Optimization	Equity	Thresholds	Uncertainty	Micro	National	Regional and Global
Cost-benefit analysis	+	+	−	+	✓	✓	✓
Risk assessment	+	+	++	++	✓	✓	✓
Multi-criteria analysis	++	+	+	+	✓	✓	
Precautionary principle[a]	+	+	++	++	✓	✓	✓
Vulnerability analysis	+	+	++	+	✓	✓	

The columns Micro, National, and Regional and Global are grouped under the heading **Scale of Application**.

[a] The precautionary principle is not strictly analogous to the other analytical and assessment methods but still can be considered a method for decision support. The precautionary principle prescribes how to bring scientific uncertainty into the decision-making process by explicitly formalizing precaution and bringing it to the forefront of the deliberations. It posits that significant actions (ranging from doing nothing to banning a potentially harmful substance or activity, for instance) may be justified when the degree of possible harm is large and irreversible.

Legend:

++ = direct application of the method by design

+ = possible application with modification or (in the case of uncertainty) the method has already been modified to handle uncertainty

− = weak but not impossible applicability with significant effort

but few of them have much to say about equity. Cost-benefit analysis can, for example, be modified to weight the interests of some people more than others. The discount rate can be viewed, in long-term analyses, as a means of weighing the welfare of future generations; and the precautionary principle can be expressed in terms of reducing the exposure of certain populations or systems whose preferential status may be the result of equity considerations. Only multicriteria analysis was designed primarily to accommodate optimization across multiple objectives with complex interactions, but this can also be adapted to consider equity and threshold issues at national and sub-national scales. Finally, the existence and significance of various thresholds for change can be explored by several tools, but only the precautionary principle was designed explicitly to address such issues.

Scenarios provide one way to cope with many aspects of uncertainty, but our limited understanding of ecological systems and human responses shrouds any individual scenario in it own characteristic uncertainty (R4.ES). Scenarios can be used to highlight the implications of alternative assumptions about critical uncertainties related to the behavior of human and ecological systems. In this way, they provide one means to cope with many aspects of uncertainty in assessing responses. The relevance, significance, and influence of scenarios ultimately depend on who is involved in their development (SG9.ES).

At the same time, though, there are a number of reasons to be cautious in the use of scenarios. First, individual scenarios represent conditional projections based on specific assumptions. Thus, to the extent that our understanding and representation of the ecological and human systems represented in the scenarios is limited, specific scenarios are characterized by their own uncertainty. Second, there is uncertainty in translating the lessons derived from scenarios developed at one scale—say, global—to the assessment of responses at other scales—say, sub-national. Third, scenarios often have hidden and hard-to-articulate assumptions. Fourth, environmental scenarios have tended to more effectively incorporate state-of-the-art natural science modeling than social science modeling.

Historically, most responses addressing ecosystem services have concentrated on the short-term benefits from increasing the productivity of provisioning services (RWG). Far less emphasis has been placed on managing regulating, cultural, and supporting ecosystem services; on management goals related to poverty alleviation and equitable distribution of benefits from ecosystem services; and on the long-term consequences of ecosystem change on the provision of services. As a result, the current management regime falls far short of the potential for meeting human needs and conserving ecosystems.

Effective management of ecosystems requires coordinated responses at multiple scales (SG9, R17.ES). Responses that are successful at a small scale are often less successful at higher levels due to constraints in legal frameworks and government institutions that prevent their success. In addition, there appear to be limits to scaling up, not only because of these higher-level constraints, but also because interventions at a local level often address only direct drivers of change rather than indirect or underlying ones. For example, a local project to improve livelihoods of communities surrounding a protected area in order to reduce pressure on it, if successful, may increase migration into buffer zones, thereby adding to pressures. Cross-scale responses may be more effective at addressing the higher-level constraints and leakage problems and simultaneously tackling regional and national as well as local-level drivers of change. Examples of successful cross-scale responses include some co-management approaches to natural resource management in fisheries and forestry and multistakeholder policy processes (R15.ES).

Active adaptive management can be a particularly valuable tool for reducing uncertainty about ecosystem management decisions (R17.4.5). The term "active" adaptive management is used here to emphasize the key characteristic of the original concept (which is frequently and inappropriately used to mean "learning by doing"): the design of management programs to test hypotheses about how components of an ecosystem function and interact and to thereby reduce uncertainty about the system more rapidly than would otherwise occur. Under an adaptive management approach, for example, a fisheries manager might intentionally set harvest levels either lower or higher than the "best estimate" in order to gain information more rapidly about the shape of the yield curve for the fishery. Given the high levels of uncertainty surrounding coupled socioecological systems, the use of active adaptive management is often warranted.

9. *What are the most important uncertainties hindering decision-making concerning ecosystems?*

The MA was unable to provide adequate scientific information to answer a number of important policy questions related to ecosystem services and human well-being. In some cases, the scientific information may well exist already but the process used and time frame available prevented either access to the needed information or its assessment. But in many cases either the data needed to answer the questions were unavailable or the knowledge of the ecological or social system was inadequate. We identify the following information gaps that, if addressed, could significantly enhance the ability of a process like the MA to answer policy-relevant questions posed by decision-makers (CWG, SWG, RWG, SGWG).

Condition and Trends

■ There are major gaps in global and national monitoring systems that result in the absence of well-documented, comparable, time-series information for many ecosystem features and that pose significant barriers in assessing condition and trends in ecosystem services. Moreover, in a number of cases, including hydrological systems, the condition of the monitoring systems that do exist is declining.

- Although for 30 years remote sensing capacity has been available that could enable rigorous global monitoring of land cover change, financial resources have not been available to process this information, and thus accurate measurements of land cover change are only available on a case study basis.
- Information on land degradation in drylands is extremely poor. Major shortcomings in the currently available assessments point to the need for a systematic global monitoring program, leading to the development of a scientifically credible, consistent baseline of the state of land degradation and desertification.
- There is little replicable data on global forest extent that can be tracked over time.
- There is no reasonably accurate global map of wetlands.

■ There are major gaps in information on nonmarketed ecosystem services, particularly regulating, cultural, and supporting services.

■ There is no complete inventory of species and limited information on the actual distributions of many important plant and animal species.

■ More information is needed concerning:

- the nature of interactions among drivers in particular regions and across scales;
- the responses of ecosystems to changes in the availability of important nutrients and carbon dioxide;

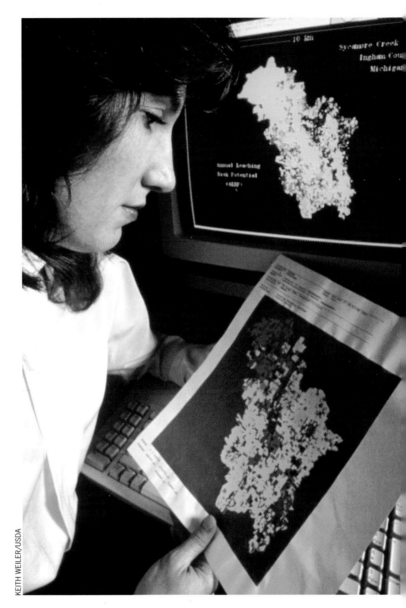

KEITH WEILER/USDA

- nonlinear changes in ecosystems, predictability of thresholds, and structural and dynamic characteristics of systems that lead to threshold and irreversible changes; and,
- quantification and prediction of the relationships between biodiversity changes and changes in ecosystem services for particular places and times.

There is limited information on the economic consequences of changes in ecosystem services at any scale and, more generally, limited information on the details of linkages between human well-being and the provision of ecosystem services, except in the case of food and water.

There are relatively few models of the relationship between ecosystem services and human well-being.

Scenarios

There is a lack of analytical and methodological approaches to explicitly nest or link scenarios developed at different geographic scales. This innovation would provide decision-makers with information that directly links local, national, regional, and global futures of ecosystem services in considerable detail.

There is limited modeling capability related to effects of changes in ecosystems on flows of ecosystem services and effects of changes in ecosystem services on changes in human well-being. Quantitative models linking ecosystem change to many ecosystem services are also needed.

Significant advances are needed in models that link ecological and social processes, and models do not yet exist for many cultural and supporting ecosystem services.

There is limited capability to incorporate adaptive responses and changes in human attitudes and behaviors in models and to incorporate critical feedbacks into quantitative models. As food supply changes, for example, so will patterns of land use, which will then feed back on ecosystem services, climate, and food supply.

There is a lack of theories and models that anticipate thresholds that, once passed, yield fundamental system changes or even system collapse.

There is limited capability of communicating to nonspecialists the complexity associated with holistic models and scenarios involving ecosystem services, in particular in relation to the abundance of nonlinearities, feedbacks, and time lags in most ecosystems.

Response Options

There is limited information on the marginal costs and benefits of alternative policy options in terms of total economic value (including nonmarketed ecosystem services).

Substantial uncertainty exists with respect to who benefits from watershed services and how changes in particular watersheds influence those services; information in both of these areas is needed in order to determine whether markets for watershed services can be a fruitful response option.

There has been little social science analysis of the effectiveness of responses on biodiversity conservation.

There is considerable uncertainty with regards to the importance people in different cultures place on cultural services, how this changes over time, and how it influences the net costs and benefits of trade-offs and decisions.

APPENDIXES

APPENDIX A
ECOSYSTEM SERVICE REPORTS

This Appendix presents some of the main findings from the Condition and Trends Working Group and the Scenarios Working Group for a selected set of ecosystem services addressed in the Millennium Ecosystem Assessment.

FOOD
PROVISIONING SERVICE

People obtain food from highly managed systems such as crops, livestock, and aquaculture and also from wild sources, including freshwater and marine capture fisheries and the harvesting of wild plants and animals (bushmeat, for example).

Condition and Trends

■ Food production more than doubled (an increase of over 160%) from 1961 to 2003 (C8.1). (See Appendix Figure A.1.) Over this period, production of cereals—the major energy component of human diets—has increased almost two and a half times, beef and sheep production increased by 40%, pork production by nearly 60%, and poultry production doubled (C8.ES).

■ Over the past 40 years, globally, intensification of cultivated systems has been the primary source (almost 80%) of increased output. But some countries, predominantly found in sub-Saharan Africa, have had persistently low levels of productivity,

and continue to rely on expansion of cultivated area. For all developing countries over the period 1961–99, expansion of harvested land contributed only 29% to growth in crop production versus the contribution of increases in yields, which amounted to 71%; in sub-Saharan Africa, however, yield increases accounted for only 34% of growth in production (C26.ES, C26.1.1).

■ Both total and per capita fish consumption have grown over the past four decades. Total fish consumption has declined somewhat in industrial countries, while it has nearly doubled in the developing world since 1973 (C8.ES).

■ Demand for fish has risen more rapidly than production, leading to increases in the real prices of most fresh and frozen fish products (C8.ES).

GLOBAL PRODUCTION, PRICES, AND UNDERNOURISHMENT

Globally, an estimated 852 million people were undernourished in 2000–02, up 37 million from the period 1997–99. Only undernourishment in developing countries is plotted in this Figure.

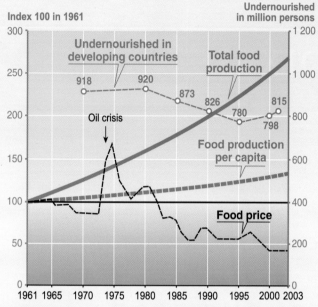

Sources: FAOSTATS, SOFI, Millennium Ecosystem Assessment

RELATIVE CHANGES IN FOOD SUPPLY (CROPS AND LIVESTOCK): INDUSTRIAL AND DEVELOPING COUNTRIES

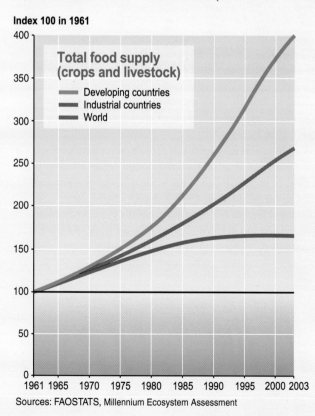

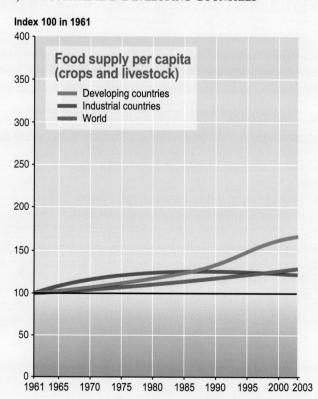

Sources: FAOSTATS, Millennium Ecosystem Assessment

■ Freshwater aquaculture is the fastest-growing food production sector. Worldwide, it has increased at an average compounded rate of 9.2% per year since 1970, compared with only 1.4% for capture fisheries and 2.8% for terrestrial farmed meat production systems (C26.3.1). Aquaculture systems now account for roughly 27% of total fish production (C8 Table 8.4).

■ The level of global output of cereals has stagnated since 1996, so grain stocks have been in decline. Although there is concern about these trends, they may reflect only a normal cycle of market adjustment (C8.2.2).

■ Although there has been some cereal price increase since 2001, prices are still some 30–40% lower than their peak in the mid-1990s (C8.2.2).

■ Current patterns of use of capture fisheries are unsustainable. Humans increased the capture of marine fish up until the 1980s by harvesting an ever-growing fraction of the available resource. Marine fish landings are now declining as a result of the overexploitation of this resource (C18.ES). Inland water fisheries, which are particularly important in providing high-quality diets for poor people, have also declined due to habitat modification, overfishing, and water withdrawals (C8.ES).

■ While traditional aquaculture is generally sustainable, an increasing share of aquaculture uses carnivorous species, and this puts increased pressure on other fisheries to provide fishmeal as feed and also exacerbates waste problems. Shrimp farming often results in severe damage to mangrove ecosystems, although some countries have taken steps to reduce these harmful impacts.

Scenarios

■ All four MA scenarios project increased total and per capita global food production by 2050 (S9). On a per capita basis, however, basic staple production stagnates or declines in the Middle East and North Africa and increases very little in sub-Saharan Africa for all four scenarios. Production shortfalls are expected to be covered through increased food imports in these regions. Agricultural land area continues to increase in developing countries under the MA scenarios, but declines in industrial countries. (See Appendix Figure A.2.)

■ Global demand for food crops (measured in tons) is projected to grow by 70–85% between 2000 and 2050 (S9.4.1).

■ Demand for both freshwater and marine fish will expand because of increasing human population and changing food preferences, and the result will be an increasing risk of a major and long-lasting decline of regional marine fisheries (*medium to high certainty*) (S9.ES).

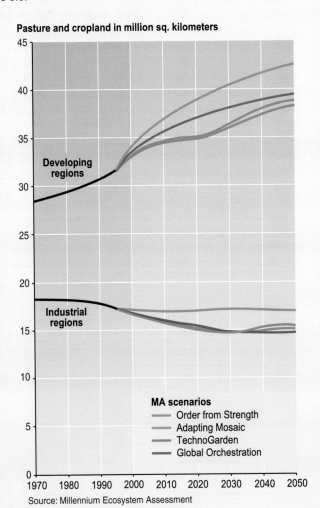

Appendix Figure A.2. CHANGES IN AGRICULTURAL LAND (PASTURE AND CROPLAND) UNDER MA SCENARIOS (S9 Fig 9.15)

Note that the total amount of pasture and cropland in 2000 plotted here is greater than the amount shown in Table 1.1 due to the fact that extensive grazing lands are included in the statistics for pasture and cropland here and not in the statistics for cultivated systems in Table 1.1.

Pasture and cropland in million sq. kilometers

Developing regions

Industrial regions

MA scenarios
Order from Strength
Adapting Mosaic
TechnoGarden
Global Orchestration

Source: Millennium Ecosystem Assessment

WATER
PROVISIONING AND SUPPORTING SERVICES

Water is both a provisioning service, since ecosystems are the source of water used by people, and a supporting service, since water is required for life on Earth and thus supports all other ecosystem processes. Forest and mountain ecosystems are associated with the largest amounts of fresh water—57% and 28% of the total runoff, respectively. These systems each provide renewable water supplies to at least 4 billion people, or two thirds of the global population. Cultivated and urban systems generate only 16% and 0.2%, respectively, of global runoff, but due to their close proximity to humans they serve from 4.5–5 billion people. Such proximity is associated with nutrient and industrial water pollution (C7.ES).

Condition and Trends

■ Recent changes to ecosystems have not significantly reduced the net amount of renewable freshwater runoff on Earth, but the fraction of that runoff used by humans has grown dramatically. Global freshwater use expanded at a mean rate of 20% per decade between 1960 and 2000, doubling over this time period (C7.ES).

■ Contemporary water withdrawal is approximately 10% of global continental runoff, although this amounts to between 40% and 50% of the continental runoff to which the majority of the global population has access during the year (C7.ES, C7.2.3).

■ Inorganic nitrogen pollution of inland waterways has increased more than twofold globally since 1960 and more than

---

Appendix Figure A.3. UNSUSTAINABLE WATER WITHDRAWALS FOR IRRIGATION (C7 Fig 7.3)

Globally, roughly 15–35% of irrigation withdrawals are estimated to be unsustainable (*low to medium certainty*) (C7.2.2). The map indicates where there is insufficient fresh water to fully satisfy irrigated crop demands. The imbalance in long-term water budgets necessitates diversion of surface water or the tapping of groundwater resources. The areas shown with moderate-to-high levels of unsustainable use occur over each continent and are known to be areas of aquifer mining or major water transfer schemes. Key: high overdraft, > 1 cubic kilometer per year; moderate, 0.1–1 cubic kilometer per year; low, 0–0.1 cubic kilometer per year. All estimates made on about 50-kilometer resolution. Though difficult to generalize, the imbalances translate into water table drawdowns >1.6 meters per year or more for the high overdraft case and <0.1 meter per year for low, assuming water deficits are met by pumping unconfined aquifers with typical dewatering potentials (specific yield = 0.2).

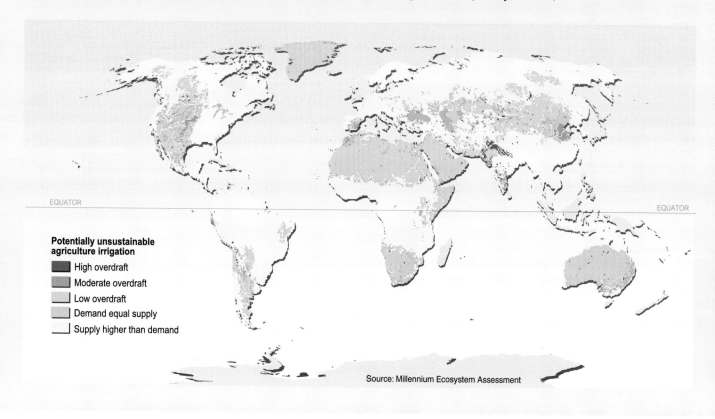

Potentially unsustainable agriculture irrigation
- High overdraft
- Moderate overdraft
- Low overdraft
- Demand equal supply
- Supply higher than demand

Source: Millennium Ecosystem Assessment

tenfold for many industrialized parts of the world (C7.ES).

■ Current patterns of human use of water are unsustainable. From 5% to possibly 25% of global freshwater use exceeds long-term accessible supplies and is met through engineered water transfers or the overdraft of groundwater supplies (*low to medium certainty*). More than 1 billion people live in areas without appreciable supplies of renewable fresh water and meet their water needs in this way (C7.ES). In North Africa and the Middle East, unsustainable use represents about a third of all water use (*low certainty*) (C7.ES).

■ Globally, 15–35% of irrigation withdrawals are estimated to be unsustainable (*low to medium certainty*) (C7.2.2). (See Appendix Figure A.3.)

Scenarios

■ Use of water is expected to grow by approximately 10% between 2000 and 2010, compared with rates of 20% per decade over the past 40 years (C7.ES).

■ Water withdrawals began to decline in many parts of the OECD at the end of the twentieth century, and with *medium certainty* will continue to decline throughout the OECD during the twenty-first century because of saturation of per capita demands, efficiency improvements, and stabilizing populations (S9.ES).

■ Water withdrawals are expected to increase greatly outside the OECD as a result of economic development and population growth. The extent of these increases is very scenario-dependent. In sub-Saharan Africa, domestic water use greatly increases and this implies (*low to medium certainty*) an increased access to fresh water. However, the technical and economic feasibility of increasing domestic water withdrawals is *very uncertain* (S9.ES).

■ Across all the MA scenarios, global water withdrawals increase between 20% and 85% between 2000 and 2050. (S9 Fig 9.35) (See Appendix Figure A.4.)

■ Global water availability increases under all MA scenarios. By 2050, global water availability increases by 5–7% (depending on the scenario), with Latin America having the smallest increase (around 2%, depending on the scenario), and the Former Soviet Union the largest (16–22%) (S9.4.5). Increasing precipitation tends to increase runoff, while warmer temperatures intensify evaporation and transpiration, which tends to decrease runoff.

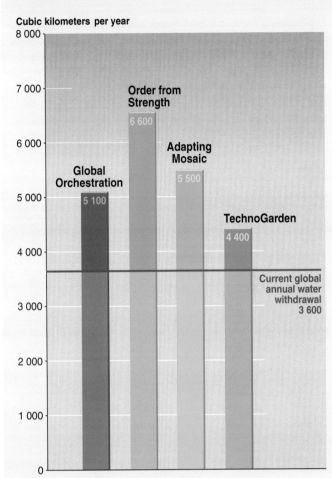

Appendix Figure A.4. Water Withdrawals in 2050 under MA Scenarios (S9 Fig 9.35)

Cubic kilometers per year

Global Orchestration 5 100
Order from Strength 6 600
Adapting Mosaic 5 500
TechnoGarden 4 400
Current global annual water withdrawal 3 600

Source: Millennium Ecosystem Assessment

TIMBER, FIBER, FUEL
PROVISIONING SERVICES

Timber is harvested from forests and plantations and used for a variety of building, manufacturing, fuel, and other needs. Forests (providing fuelwood and charcoal), agricultural crops, and manure all serve as sources of biomass energy. A wide variety of crops and livestock are used for fiber production. Cotton, flax, hemp, and jute are generally produced from agricultural systems, while sisal is produced from the leaves of Agave cactus. Silk is produced by silkworms fed the leaves of the mulberry tree, grown in an orchard-like culture, and wool is produced by sheep, goats, alpaca, and other animals.

Appendix Figure A.5. CHANGES IN FOREST AREA UNDER MA SCENARIOS (S9 Fig 9.15)

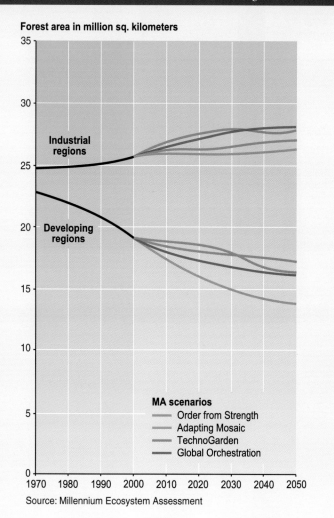

Forest area in million sq. kilometers

MA scenarios
- Order from Strength
- Adapting Mosaic
- TechnoGarden
- Global Orchestration

Source: Millennium Ecosystem Assessment

Condition and Trends

■ Global timber harvests increased by 60% since 1960, and wood pulp production increased slightly less than threefold over this same time (C9.ES, C9 Table 9.5). Rates of growth in harvests have slowed in recent years.

■ Fuelwood is the primary source of energy for heating and cooking for some 2.6 billion people, and 55% of global wood consumption is for fuelwood (C9.ES). Although they account for less than 7% of world energy use, fuelwood and charcoal provide 40% of energy use in Africa and 10% in Latin America (C9.4).

■ Global consumption of fuelwood appears to have peaked in the 1990s and is now believed to be slowly declining as a result of switching to alternate fuels and, to a lesser degree, more-efficient biomass energy technologies. In contrast, global consumption of charcoal appears to have doubled between 1975 and 2000, largely as a result of continuing population shifts toward urban areas (C9.4.1).

■ Localized fuelwood shortages in Africa impose burdens on people who depend on fuelwood for home heating and cooking (SG3.4). The impact on people may be high prices in urban areas or lengthy and arduous travel to collect wood in rural areas.

■ Among agricultural fibers, global cotton production has doubled and silk production has tripled since 1961 (C9.ES). Despite this doubling of production, the land area on which cotton is harvested has stayed virtually the same. Production of flax, wool, hemp, jute, and sisal has declined. For example, competition from synthetic fabrics has contributed to a reduction in the demand for wool in recent decades; wool production declined 16% between 1980 and 2000 (C9.5.3).

Scenarios

■ Plantations are likely to provide an increasing proportion of timber products in the future (C9.ES). In 2000, plantations were 5% of the global forest cover, but they provided some 35% of harvested roundwood, an amount anticipated to increase to 44% by 2020. The most rapid expansion will occur in the mid-latitudes, where yields are higher and production costs lower.

■ Under the MA scenarios, forest area increases in industrial regions and decreases in developing ones between 1970 and 2050. In one scenario (*Order from Strength*), the rate of forest loss increases from the historic rate (of about 0.4% annually between 1970 and 1995) to 0.6%. In *Global Orchestration* and *Adapting Mosaic*, the rate of loss continues at the historic rate. Forest loss in *TechnoGarden* decreases in the first decades of the scenario period, but over the whole period is near the historic rate because the use of biofuels increases as part of climate change policies, leading to further pressure on forest area. (See Appendix Figure A.5.) (For particular ecosystems, such as tropical forests, deforestation rates might be higher than average.)

BIOCHEMICALS AND GENETIC RESOURCES
PROVISIONING SERVICES

A wide variety of species—microbial, plant, and animal—and their genes contribute to commercial products in such industries as pharmaceuticals, botanical medicines, crop protection, cosmetics, horticulture, agricultural seeds, environmental monitoring and a variety of manufacturing and construction sectors.

Condition and Trends

■ Biodiversity is in increasing demand as a source of commercial material. An overview of the industries involved, trends in the use of biodiversity, and the types of social and commercial benefits is provided in Appendix Table A.1. Appendix Table A.2 is a partial list of compounds derived from natural sources approved for marketing within the pharmaceutical industry in the 1990s.

Scenarios

■ Market trends vary widely according to the industry and country involved, but many bioprospecting activities and revenues are expected to increase over the next decades. Several major new industries, such as bioremediation and biomimetics, are well established and appear set to increase, while others have a less certain future. The current economic climate suggests that pharmaceutical bioprospecting will increase, especially as new methods that use evolutionary and ecological knowledge enhance productivity (C10.ES).

Appendix Table A.1. A Summary of Status and Trends in Major Bioprospecting Industries (C10 Table 10.8)

Industry	Current Involvement in Bioprospecting	Expected Trend in Bioprospecting	Social Benefits	Commercial Benefits	Biodiversity Resources
Pharmaceutical	tends to be cyclical	cyclical, possible increase	human health, employment	+++	P,A,M
Botanical	high	increase	human health, employment	+++	mostly P,A,M
Cosmetics and natural personal care	high	increase	human health and well-being	+++	P,A,M
Bioremediation	variable	increase	environmental health	++	mostly M
Crop protection and biological control	high	increase	food supply, environmental health	+++	P,A,M
Biomimetics	variable	variable, increasing?	various	++	P,A,M
Biomonitoring	variable	increase	environmental health	+	P,A,M
Horticulture and seed industry	low	steady	human well-being, food supply	+++	P
Ecological restoration	medium	increase	environmental health	++	P,A,M

Legend: +++ = billion dollar, ++ = million dollar, + profitable but amounts vary
P= plants, A = animals, M= microorganisms

Appendix Table A.2. SOME COMPOUNDS FROM NATURAL SOURCES (PURE NATURAL PRODUCTS, SEMI-SYNTHETIC MODIFICATIONS, OR THE PHARMACOPHORE IS FROM A NATURAL PRODUCT) APPROVED FOR MARKETING IN THE 1990S, IN THE UNITED STATES AND ELSEWHERE (C10 Table 10.2)

Generic	Brand Name	Developer
In the United States and elsewhere		
Cladribine	Leustatin	Johnson & Johnson (Ortho Biotech)
Docetaxel	Taxotere	Rhône-Poulenc Rorer
Fludarabine	Fludara	Berlex
Idarubicin	Idamycin	Pharmacia & Upjohn
Irinotecan	Camptosar	Yakult Haisha
Paclitaxel	Taxol	Bristol-Myers Squibb
Pegaspargase	Oncospar	Rhône-Poulenc
Pentostatin	Nipent	Parke-Davis
Topotecan	Hycamtin	SmithKline Beecham
Vinorelbine	Navelbine	Lilly
Only outside the United States		
Bisantrene		Wyeth Ayerst
Cytarabine ocfosfate		Yamasa
Formestane		Ciba-Geigy
Interferon, gamma-la		Siu Valy
Miltefosine		Acta Medica
Porfimer sodium		Quadra Logic
Sorbuzoxane		Zeuyaku Kogyo
Zinostatin		Yamamouchi

CLIMATE REGULATION
REGULATING SERVICES

Ecosystems, both natural and managed, exert a strong influence on climate and air quality as sources and sinks of pollutants, reactive gases, greenhouse gases, and aerosols and due to physical properties that affect heat fluxes and water fluxes (precipitation). Ecosystems can affect climate in the following ways: warming (as sources of greenhouse gases, for instance, or forests with lower albedo than bare snow); cooling (as sinks of greenhouse gas, sources of some aerosol that reflect solar radiation, and evapotranspiration, for example); and by altering water redistribution/recycling and regional rainfall patterns (through evapotranspiration, for instance, or cloud condensation nuclei).

Condition and Trends

■ Changes in ecosystems have made a large contribution to historical changes in radiative forcing from 1750 to the present mainly due to deforestation, fertilizer use, and agricultural practices (C13.ES). (See Appendix Figure A.6.) Ecosystem changes account for about 10–30% of the radiative forcing of CO_2 since 1750 and a large proportion of the radiative forcing due to CH_4 and N_2O. Ecosystems are currently a net sink for CO_2 and tropospheric ozone, while they remain a net source of CH_4 and N_2O. Future management of ecosystems has the potential to modify concentrations of a number of greenhouse gases, although this potential is likely to be small in comparison to IPCC scenarios of fossil fuel emissions over the next century (*high certainty*). Ecosystems influence the main anthropogenic greenhouse gases as follows:

- Carbon dioxide: About 40% of the historical emissions (over the last two centuries), and about 20% of current CO_2 emissions (in the 1990s), originated from changes in land use and land management, primarily deforestation. Terrestrial ecosystems were a sink for about a third of cumulative historical emissions and a third of total emissions in the 1990s (energy plus land use). The sink may be explained partially by afforestation, reforestation, and forest management in North America, Europe, China, and other regions and partially by the fertilizing effects of N deposition and increasing atmospheric CO_2. Terrestrial ecosystems were on average a net source of CO_2 during the nineteenth and early twentieth centuries and became a net sink sometime around the middle of the last century (*high certainty*). The net impact of ocean biology changes on global CO_2 fluxes is unknown.

- Methane: Natural processes in wetland ecosystems account for about 25–30% of current methane emissions, and about 30% of emissions are due to agriculture (ruminant animals and rice paddies).

- Nitrous oxide: Ecosystem sources account for about 90% of current N_2O emissions, with 35% of emissions from agricultural systems, primarily driven by fertilizer use.

- Tropospheric ozone: Dry deposition in ecosystems accounts for about half the tropospheric ozone sink. Several gases emitted by ecosystems, primarily due to biomass burning, act as precursors for tropospheric ozone formation (NO_X, volatile organic compounds, CO, CH_4). The net global effect of ecosystems is as a sink for tropospheric O_3.

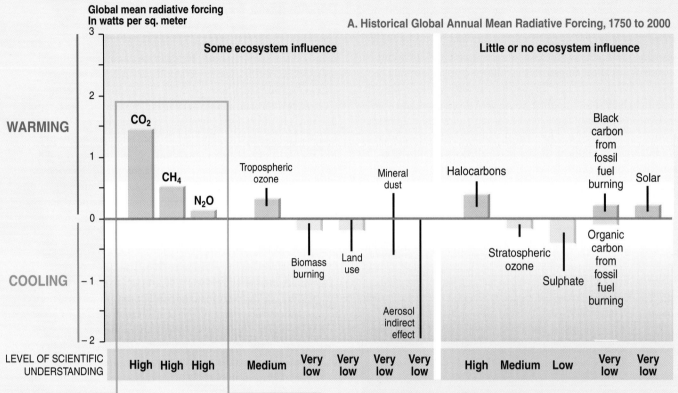

Global mean radiative forcing
In watts per sq. meter

A. Historical Global Annual Mean Radiative Forcing, 1750 to 2000

Some ecosystem influence | Little or no ecosystem influence

WARMING

COOLING

CO₂ · CH₄ · N₂O · Tropospheric ozone · Biomass burning · Land use · Mineral dust · Aerosol indirect effect · Halocarbons · Stratospheric ozone · Sulphate · Black carbon from fossil fuel burning · Organic carbon from fossil fuel burning · Solar

| LEVEL OF SCIENTIFIC UNDERSTANDING | High | High | High | Medium | Very low | Very low | Very low | Very low | High | Medium | Low | Very low | Very low |

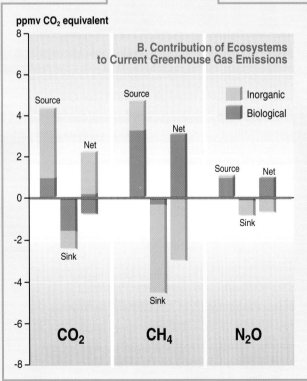

ppmv CO₂ equivalent

B. Contribution of Ecosystems to Current Greenhouse Gas Emissions

Inorganic
Biological

CO₂ · CH₄ · N₂O

Source · Sink · Net

NB: The height of a bar indicates a best estimate of the forcing, and the accompanying vertical black line a likely range of values. Where no bar is present, the vertical line only indicates the range in best estimates with no likelihood.

Sources: IPCC, Millennium Ecosystem Assessment

Figure A is the radiative forcing caused by changes in atmospheric composition, alteration in land surface reflectance (albedo), and variation in the output of the sun for the year 2000 relative to conditions in 1750. The height of the bar represents a best estimate, and the accompanying vertical line a likely range of values. Factors with a significant ecosystem influence are separated from those without one. The indirect effect of aerosols shown is their effect on cloud droplet size and number, not cloud lifetime.

Figure B is the relative contribution of ecosystems to sources, sinks, and net changes in three main greenhouse gases. These can be compared with each other by conversion into CO₂-equivalent values, based on the global warming potential (radiative impact times atmospheric lifetime) of the different gases. For CH₄ and N₂O, a 100-year time scale was assumed; a short time scale would increase the relative value compared with CO₂ and a longer time scale would reduce it. Ecosystems are also a net sink for tropospheric ozone, but it is difficult to calculate emissions in CO₂-equivalent values.

■ During much of the past century, most cropping systems have undergone a steady net loss of soil organic matter. However, with the steady increase in crop yields, which increases crop biomass and the amount of residue returned to the soil, and with the adoption of conservation tillage and no-till cropping systems, net carbon sequestration is estimated to occur in the maize-soybean systems of North America and in some continuous irrigated lowland rice systems. Agriculture accounts for 44% of anthropogenic methane emissions and about 70% of anthropogenic nitrous oxide gases, mainly from the conversion of new land to agriculture and nitrogen fertilizer use (C26.2.6).

■ Terrestrial and marine plants fix atmospheric CO_2 and return it via respiration. In the ocean, some of the carbon sinks in the form of dead organisms, particles, and dissolved organic carbon, a small amount of which remains in sediments; the rest is respired at depth and eventually recirculated to the surface (the "biological pump"). The biological pump acts as a net sink for CO_2 by increasing its concentration at depth, where it is isolated from the atmosphere for decades to centuries, causing the concentration of CO_2 in the atmosphere to be about 200 parts per million lower than it would be in the absence of life (C13.2.1). On the land large amounts of carbon fixed by plants are stored in soil organic matter.

■ Land cover changes since 1750 have increased the reflectivity to solar radiation (albedo) of the land surface (*medium certainty*), partially offsetting the warming effect of associated CO_2 emissions (C13.ES). Deforestation and desertification in the tropics and sub-tropics leads to a reduction in regional rainfall (*high certainty*). Biophysical effects need to be accounted for in the assessment of options for climate change mitigation. For example, the warming effect of reforestation in seasonally snow-covered regions due to albedo decrease is likely to exceed the cooling effect of additional carbon storage in biomass. Biophysical effects of ecosystem changes on regional climate patterns depend on geographical location and season. With *high certainty*:

- Deforestation in seasonally snow-covered regions leads to regional cooling of the land surface during the snow season due to increase in surface albedo, and it leads to warming during the summer due to reduction in evapotranspiration.
- Large-scale tropical deforestation (hundreds of square kilometers) reduces regional rainfall, primarily due to decreased evapotranspiration.
- Desertification in the tropics and sub-tropics leads to decrease in regional precipitation due to reduced evapotranspiration and increased surface albedo.

Scenarios

■ The future contribution of terrestrial ecosystems to the regulation of climate is uncertain. Currently, the biosphere is a net sink of carbon, absorbing about 1–2 gigatons of carbon per year, or approximately 20% of fossil fuel emissions. It is very likely that the future of this service will be greatly affected by expected land use change. In addition, a higher atmospheric CO_2 concentration is expected to enhance net productivity, but this does not necessarily lead to an increase in the carbon sink. The limited understanding of soil respiration processes generates uncertainty about the future of the carbon sink. There is *medium certainty* that climate change will increase terrestrial fluxes of CO_2 and CH_4 in some regions (such as in Arctic tundras) (S9.ES).

DISEASE REGULATION
REGULATING SERVICES

The availability of many ecosystem services, such as food, water, and fuel, can profoundly influence human health (R16). Here, we consider a much narrower service provided by ecosystems related to human health: the role of ecosystems in regulating infectious disease. Ecosystem changes have played an important role in the emergence or resurgence of infectious diseases. (See Appendix Table A.3.) Ecosystem modifications associated with developments such as dam building and the expansion of agricultural irrigation, for example, have sometimes increased the local incidence of infectious diseases such as malaria, schistosomiasis, and arbovirus infections, especially in the tropics. Other modifications to ecosystems have served to reduce the incidence of infectious disease.

Condition and Trends

■ Infectious diseases still account for close to one quarter of the global burden of disease. Major tropical diseases, particularly malaria, meningitis, leishmaniasis, dengue, Japanese encephalitis, African trypanosomiasis, Chagas disease, schistosomiasis, filariasis, and diarrheal diseases still infect millions of people throughout the world (*very certain*) (C14.ES).

■ The prevalence of the following infectious diseases is particularly strongly influenced by ecological change: malaria across most ecological systems; schistosomiasis, lymphatic filariasis, and Japanese encephalitis in cultivated and inland water systems in the tropics; dengue fever in tropical urban centers; leishmaniasis and Chagas disease in forest and dryland systems; meningitis in the Sahel; cholera in coastal, freshwater, and urban systems; and West Nile virus and Lyme disease in urban and suburban systems of Europe and North America (*high certainty*) (C14.ES).

■ Various changes to ecosystems can affect disease incidence through a variety of mechanisms. Disease/ecosystem relationships that best exemplify these biological mechanisms include the following examples (C14.ES):

■ Dams and irrigation canals provide ideal habitat for snails that serve as the intermediate reservoir host species for schistosomiasis; irrigated rice fields increase in the extent of mosquito-breeding surface, increasing the chance of transmission of mosquito-borne malaria, lymphatic filariasis, Japanese encephalitis, and Rift Valley fever.

■ Deforestation has increased the risk of malaria in Africa and South America by increasing habitat suitable for malaria-transmitting mosquitoes.

■ Natural systems with preserved structure and characteristics generally resist the introduction of invasive human and animal pathogens brought by human migration and settlement. This seems to be the case for cholera, kala-azar, and schistosomiasis, which did not become established in the Amazonian forest ecosystem (*medium certainty*).

■ Uncontrolled urbanization in the forest ecosystem has been associated with mosquito-borne viruses (arboviruses) in the Amazon and with lymphatic filariasis in Africa. Tropical urban areas with poor water supply systems and lack of shelter promote transmission of dengue fever.

■ There is evidence that habitat fragmentation, with subsequent biodiversity loss, increases the prevalence in ticks of the bacteria that causes Lyme disease in North America (*medium certainty*).

■ Zoonotic pathogens (defined by their natural life cycle in animals) are a significant cause of both historical (such as HIV and tuberculosis) and newly emerging infectious diseases affecting humans (such as SARS, West Nile virus, and Hendra virus). In addition, zoonotic pathogens can cause high case-fatality rates and are difficult to vaccinate against, since the primary reservoir hosts are nonhumans.

■ Intensive livestock agriculture that uses subtherapeutic doses of antibiotics has led to the emergence of antibiotic-resistant strains of *Salmonella*, *Campylobacter*, and *Escherichia coli bacteria*. Overcrowded and mixed livestock practices, as well as the trade in bushmeat, can facilitate interspecies host transfer of disease agents, leading to dangerous novel pathogens such as SARS and new strains of influenza.

Scenarios

■ Tropical developing countries are more likely to be affected in the future due to the greater exposure of people in these countries to vectors of infectious disease transmission. Such populations have a scarcity of resources to respond to disease and to plan environmental modifications associated with economic activities (*high certainty*). However, international trade and transport leave no country entirely unaffected (S11).

■ The health consequences under the MA scenarios related to changes in the disease regulation service of ecosystems vary widely, with some scenarios showing improving conditions and others declining conditions (S11).

Disease	Cases Per Year[a]	Disability-adjusted Life Years[b] (thousands)	(Proximate) Emergence Mechanism	(Ultimate) Emergence Driver	Geographical Distribution	Expected Variation from Ecological Change	Confidence Level
Marlaria	350 m	46,486	niche invasion; vector expansion	deforestation; water projects	tropical (America, Asia, and Africa)	++++	+++
Dengue fever	80 m	616	vector expansion	urbanization; poor housing conditions	tropical	+++	++
HIV	42 m	84,458	host transfer	forest encroachment; bushmeat hunting; human behavior	global	+	++
Leishmaniasis	12 m	2090	host transfer; habitat alteration	deforestation; agricultural development	tropical Americas; Europe and Middle East	++++	+++
Lyme disease	23,763 (US 2002)		depletion of predators; biodiversity loss; reservoir expansion	habitat fragmentation	North America and Europe	++	++
Chagas disease	16–18 m	667	habitat alteration	deforestation; urban sprawl and encroachment	Americas	++	+++
Japanese encephalitis	30–50,000	709	vector expansion	irrigated rice fields	Southeast Asia	+++	+++
West Nile virus and other encephalitides	–	–			Americas and Eurasia	++	+
Guanarito; Junin, Machupo	–	–	biodiversity loss; reservoir expansion	monoculture in agriculture after deforestation	South America	++	+++
Oropouche/ Mayaro o virus in Brazil	–	–	vector expansion	forest encroachment; urbanization	South America	+++	+++
Hantavirus	–	–	variations in population density of natural food sources	climate variability		++	++
Rabies	–	–	biodiversity loss; altered host selection	deforestation and mining	tropical	++	++
Schistosomiasis	120 m	1,702	intermediate host expansion	dam building; irrigation	America, Africa, and Asia	++++	++++
Leptospirosis	–	–			global (tropical)	++	+++
Cholera	†	¥	sea surface temperature rising	climate variability and change	global (tropical)	+++	++
Cryptosporidiosis	†	¥	contamination by oocystes	poor watershed management where livestock exist	global	+++	++++

(*continued on page 116*)

Disease	Cases Per Year[a]	Disability-adjusted Life Years[b] (thousands)	(Proximate) Emergence Mechanism	(Ultimate) Emergence Driver	Geographical Distribution	Expected Variation from Ecological Change	Confidence Level
Meningitis		6,192	dust storms	desertification	Saharan Africa	++	++
Coccidioido-mycosis	–	–	disturbing soils	climate variability	global	++	+++
Lymphatic Filariasis	120 m	5,777			tropical America and Africa	+	+++
Trypanosomiasis	30–500,000	1,525			Africa		
Onchocerciasis	18 m	484			Africa and tropical America	++	+++
Rift Valley Fever			heavy rains	climate variability and change	Africa		
Nipah/Hendra viruses			niche invasion	industrial food production; deforestation; climate abnormalities	Australia and Southeast Asia	+++	+
Salmonellosis			niche invasion	antibiotic resistance from using antibiotics in animal feed			
Ebola			forest encroachment; bushmeat hunting				
BSE			host transfer	intensive livestock farming			
SARS			host transfer	intensive livestock operations mixing wild and domestic animals			

[a] m = millions

[b] Disability-adjusted life years: years of healthy life lost—a measure of disease burden for the gap between actual health of a population compared with an ideal situation where everyone lives in full health into old age.

† and ¥ Diarheal diseases (aggregated) deaths and DALYs respectively: 1,798 X 1,000 cases and 61,966 X 1,000 DALYs

Legend: + = low; ++ = moderate; +++ high; ++++ = very high

WASTE TREATMENT
REGULATING SERVICES

Because the characteristics of both wastes and receiving ecosystems vary, environments vary in their ability to absorb wastes and to detoxify, process, and sequester them. Some contaminants (such as metals and salts) cannot be converted to harmless materials, but others (organic chemicals and pathogens, for example) can be degraded to harmless components. Nevertheless, these materials may be released to the environment fast enough to modify ecosystem functioning significantly. Some materials (such as nutrient fertilizers and organic matter) are normal components of organism metabolism and ecosystem processes. Nevertheless, loading rates of these materials may occur fast enough to modify and impair ecosystem function significantly.

Condition and Trends

■ The problems associated with wastes and contaminants are in general growing. Some wastes—sewage, for instance—are produced in nearly direct proportion to population size. Other types of wastes and contaminants reflect the affluence of society. An affluent society uses and generates a larger volume of waste-producing materials such as domestic trash and home-use chemicals (C15.ES).

■ Where there is significant economic development, loadings of certain wastes are expected to increase faster than population growth. The generation of some wastes (industrial waste, for example) does not necessarily increase with population or development state. These wastes may often be reduced through regulation aimed to encourage producers to clean discharges or to seek alternate manufacturing processes (C15.ES).

■ In developing countries, 90–95% of all sewage and 70% of industrial wastes are dumped untreated into surface water (C7.4.5). Regional patterns of processing nitrogen loads in freshwater ecosystems provide a clear example of the overloading of the waste processing service of ecosystems.

■ Aquatic ecosystems "cleanse" on average 80% of their global incident nitrogen loading but this intrinsic self-purification capacity of these ecosystems varies widely and is not unlimited (C7.2.5).

■ Severe deterioration in the quality of fresh water is magnified in cultivated and urban systems (high use, high pollution sources) and in dryland systems (high demand for flow regulation, absence of dilution potential) (C7.ES).

Scenarios

■ It is neither possible nor appropriate to attempt to state whether the intrinsic waste detoxification capabilities of the planet as a whole will increase or decrease with a changing environment. The detoxification capabilities of individual locations may change with changing conditions (such as changes in soil moisture levels). At high waste-loading rates, however, the intrinsic capability of environments is overwhelmed, such that wastes will build up in the environment to the detriment of human well-being and a loss of biodiversity (C15.ES).

■ The service of water purification could be either enhanced or degraded in both developing and industrial countries under the MA Scenarios (S9.5.4). Within industrial countries, the dilution capacity of most rivers increases because higher precipitation leads to increases in runoff in most river basins. Wetland areas decrease because of the expansion of population and agricultural land. Wastewater flows increase, but in some scenarios the wealth of the North enables it to repair breakdowns in water purification as they occur. Within developing countries, the pace of ecosystem degradation, the overtaxing of ecosystems by high waste loads, and the decline of wetland area because of the expansion of population and agricultural land tend to drive a deterioration of water purification in two scenarios. The *Adapting Mosaic* scenario, however, could lead to some gains in water purification even in developing countries, and the *TechnoGarden* scenario would also result in gains.

Natural Hazard Regulation
Regulating Services

Ecosystems play important roles in modulating the effects of extreme events on human systems. Ecosystems affect both the probability and severity of events, and they modulate the effects of extreme events. Soils store large amounts of water, facilitate transfer of surface water to groundwater, and prevent or reduce flooding. Barrier beaches, wetlands, and lakes attenuate floods by absorbing runoff peaks and storm surges.

Condition and Trends

■ Humans are increasingly occupying regions and localities that are exposed to extreme events, (such as on coasts and floodplains or close to fuelwood plantations). These actions are exacerbating human vulnerability to extreme events, such as the December 2004 tsunami in the Indian Ocean. Many measures of human vulnerability show a general increase due to growing poverty, mainly in developing countries (C16.ES).

■ Roughly 17% of all the urban land in the United States is located in the 100-year flood zone. Likewise, in Japan about 50% of the population lives on floodplains, which cover only 10% of the land area. In Bangladesh, the percentage of flood-prone areas is much higher and inundation of more than half of the country is not uncommon. For example, about two thirds of the country was inundated in the 1998 flood (C16.2.2).

■ Many of the available datasets on extreme events show that impacts are increasing in many regions around the world. From 1992 to 2001, floods were the most frequent natural disaster (43% of 2,257 disasters), and they killed 96,507 people and affected more than 1.2 billion people over the decade. Annual economic losses from extreme events increased tenfold from the 1950s to the 1990s (C16.ES).

■ The loss of ecosystems such as wetlands and mangroves has significantly reduced natural mechanisms of protection from natural hazards. For example, forested riparian wetlands adjacent to the Mississippi River in the United States during presettlement times had the capacity to store about 60 days of river discharge. With the removal of wetlands through canalization, leveeing, and draining, the remaining wetlands have a storage capacity of less than 12 days discharge—an 80% reduction of flood storage capacity (C16.1.1).

■ The number of floods and fires increased significantly on all continents over the past 60 years. (See Appendix Figures A.7 and A.8.)

■ Within industrial countries, the area burned by fires is declining but the number of major fires is increasing. In the United States, for example, the area burned has declined by more than 90% since 1930, while in Sweden the area burned annually fell from about 12,000 hectares in 1876 to about 400 hectares in 1989. In North America, however, the number of fire "disasters"—10 or more people reportedly killed, 100 people reportedly affected, a declared state of emergency, and a call for international assistance—increased from about 10 in the 1980s to about 45 during the 1990s (C16.2.2).

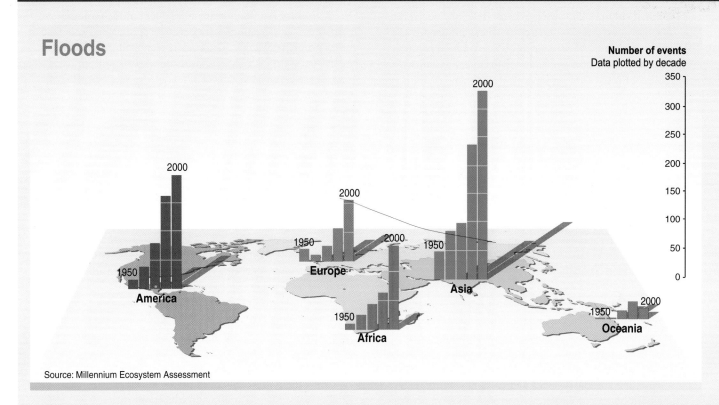

Source: Millennium Ecosystem Assessment

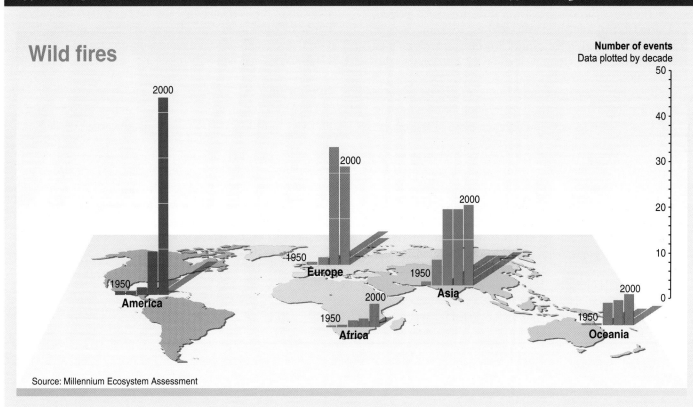

Source: Millennium Ecosystem Assessment

CULTURAL SERVICES

Human cultures, knowledge systems, religions, social interactions, and amenity services have been influenced and shaped by the nature of ecosystems. At the same time, humankind has influenced and shaped its environment to enhance the availability of certain valued services. Recognizing that it is not possible to fully separate the different spiritual, intellectual, and physical links between human cultures and ecosystems, the MA assessed six main types of cultural and amenity services provided by ecosystems: cultural diversity and identity; cultural landscapes and heritage values; spiritual services; inspiration (such as for arts and folklore); aesthetics; and recreation and tourism. Because global aggregated information on the condition of cultural services was limited (with the partial exception of recreational and tourism benefits), the section below draws significantly on information in the MA sub-global assessments.

Condition and Trends

■ Transformation of once diverse ecosystems into relatively more similar cultivated landscapes, combined with social and economic changes including rapid urbanization, breakdown of extended families, loss of traditional institutions, easier and cheaper transportation, and growing economic and social "globalization," has significantly weakened the linkages between ecosystems and cultural diversity and cultural identity (C17.2.1). Throughout human evolution, human societies have developed in close interaction with the natural environment, which has shaped their cultural identity, value systems, and language.

■ The loss of particular ecosystem attributes (sacred species or sacred forests), combined with social and economic changes, can sometimes weaken the spiritual benefits people obtain from ecosystems in many parts of the world (C17.2.3). On the other hand, under some circumstances (such as where ecosystem attributes are causing significant threats to people) the loss of some attributes may enhance spiritual appreciation for what remains.

■ People across cultures and regions express, in general, an aesthetic preference for natural environments over urban or built ones; the conversion and degradation of relatively natural environments has diminished these benefits. Ecosystems continue to inspire arts, songs, drama, dance, design, and fashion, although the stories told through such media are different from those told historically (C17.2.5).

■ Recreation and tourism uses of ecosystems are growing, due to growing populations, greater leisure time available among wealthy populations, and greater infrastructure development to support recreational activities and tourism. Nature travel increased at an estimated rate of 10–30% annually in the early 1990s, and in 1997 nature tourism accounted for approximately 20% of total international travel (C17.2.6). Tourism is now the primary economic development strategy for a number of developing countries.

■ Tourism is an important component of the economies of many of the MA sub-global assessment study areas, and at all scales most assessment stakeholders requested its inclusion. In contrast, spiritual, religious, recreational, and educational services tended to be assessed only at a fine scale in small local studies, typically because the data required for these assessments are not available at a broad scale and because of the culture-specific, intangible, and sometimes sensitive nature of these services (SG8.3).

■ Within the MA sub-global assessments, cultural services of tourism and recreation were generally in a good condition and growing, although some assessments expressed concerns about tourist activities potentially reducing the capacity of ecosystems to provide this cultural service (SG8.3).

■ In contrast, within the MA sub-global assessments local-scale services of a spiritual nature are of a variable condition, typically either collapsing or being revived, depending on policies, interventions, and context-specific factors such as changes in leadership (SG8.3). Spiritual values were found to act as strong incentives for ecosystem conservation in sub-global assessments in Peru, Costa Rica, India, and some parts of Southern Africa. Educational services of ecosystems assessed in Sweden, São Paulo, and Portugal are all increasing due to growing levels of awareness of the value and benefits of, and thus the demand for, environmental education.

■ While provisioning services such as water, medicinal plants, fuelwood, and food are very important, spiritual and sacred elements in the local landscape also have a very specific and important value to local people across all the assessments. In several cases, spiritual values coincided with other values, such as biodiversity, water supply, biomedicines, and fuel (SG11.3).

Scenarios

■ The MA Scenarios project changes in cultural services based only on a qualitative analyses due to the absence of suitable quantitative models. Cultural services increase in some scenarios and decline in others. Generally, cultural services decline moderately in *Global Orchestration* and strongly in *Order from Strength*, driven in both cases by lack of personal experience with nature and lower cultural diversity. Lower cultural diversity also drives a decline in cultural services in the *TechnoGarden* scenario. On the other hand, cultural services increase in *Adapting Mosaic*, due in part to the increase in knowledge systems and cultural diversity (S9.7).

Nutrient Cycling
Supporting Services

An adequate and balanced supply of elements necessary for life, provided through the ecological processes of nutrient cycling, underpins all other ecosystem services. The cycles of several key nutrients have been substantially altered by human activities over the past two centuries, with important positive and negative consequences for a range of other ecosystem services and for human well-being. Nutrients are mineral elements such as nitrogen, phosphorus, and potassium that are essential as raw materials for organism growth and development. Ecosystems regulate the flows and concentrations of nutrients through a number of complex processes that allow these elements to be extracted from their mineral sources (atmosphere, hydrosphere, or lithosphere) or recycled from dead organisms. This service is supported by a diversity of different species.

Condition and Trends

■ The capacity of terrestrial ecosystems to absorb and retain the nutrients supplied to them either as fertilizers or from the deposition of airborne nitrogen and sulfur has been undermined by the radical simplification of ecosystems into large-scale, low-diversity agricultural landscapes. Excess nutrients leak into the groundwater, rivers, and lakes and are transported to the coast. Treated and untreated sewage released from urban areas adds to the load (C.SDM).

■ In preindustrial times, the annual flux of nitrogen from the atmosphere to the land and aquatic ecosystems was roughly 110–210 teragrams of nitrogen a year. Human activity contributes an additional 165 teragams or so of nitrogen per year, roughly doubling the rate of creation of reactive N on the land surfaces of Earth (R9.2). (See Appendix Figure A.9.)

Appendix Figure A.9. Contrast between Contemporary and Pre-disturbance Transports of Total Nitrogen through Inland Aquatic Systems Resulting from Anthropogenic Acceleration of This Nutrient Cycle (C7 Fig 7.5)

While the peculiarities of individual pollutants, rivers, and governance define the specific character of water pollution, the general patterns observed for nitrogen are representative of anthropogenic changes to the transport of waterborne constituents. Elevated contemporary loadings to one part of the system (such as croplands) often reverberate to other parts of the system (to coastal zones, for example), exceeding the capacity of natural systems to assimilate additional constituents.

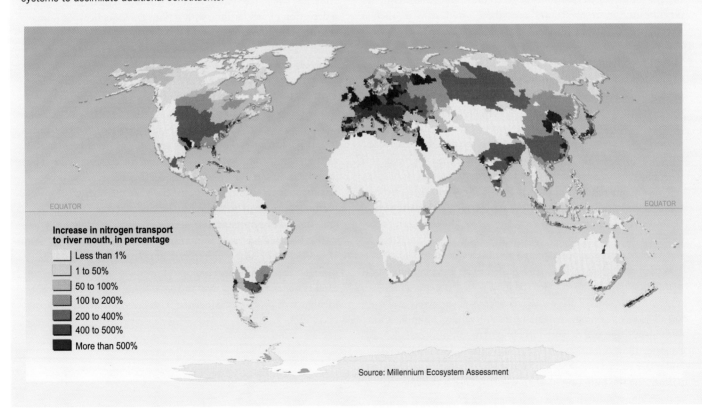

Increase in nitrogen transport to river mouth, in percentage

- Less than 1%
- 1 to 50%
- 50 to 100%
- 100 to 200%
- 200 to 400%
- 400 to 500%
- More than 500%

Source: Millennium Ecosystem Assessment

■ The N accumulation on land and in waters has permitted a large increase in food production in some countries, but at the cost of increased emissions of greenhouse gases and frequent deterioration in freshwater and coastal ecosystem services, such as water quality, fisheries, and amenity values (C12.ES).

■ Phosphorus is also accumulating in ecosystems at a rate of 10.5–15.5 teragrams per year, compared with a preindustrial rate of 1–6 teragrams per year, mainly as a result of the use of phosphorus (obtained through mining) in agriculture. Most of this accumulation is in soils. If these soils erode into freshwater systems, deterioration of ecosystem services can result. This tendency is likely to spread and worsen over the next decades, since large amounts of P have been accumulated on land and their transport to water systems is slow and difficult to prevent (C12.ES).

■ Sulfur emissions have been progressively reduced in Europe and North America but not yet in the emerging industrial areas of the world: China, India, South Africa, and the southern parts of South America. A global assessment of acid deposition threats suggests that tropical ecosystems are at high risk (C12.ES).

■ Human actions at all scales required to feed the current world population have increased the "leakiness" of ecosystems with respect to nutrients. Tillage often damages soil structure, and the loss of biodiversity may increase nutrient leaching. Simplification of the landscape and destruction of riparian forests, wetlands, and estuaries allow unbuffered flows of nutrients between terrestrial and water ecosystems. Specific forms of biodiversity are critical to performing the buffering mechanisms that ensure the efficient use and cycling of nutrients in ecosystems (C12.ES).

■ In contrast to these issues associated with nutrient oversupply, there remain large parts of Earth, notably in Africa and Latin America, where harvesting without nutrient replacement has led to a depletion of soil fertility, with serious consequences for human nutrition and the environment (C12.ES).

Scenarios

■ Recent scenario studies that include projections of nitrogen fertilizer use indicate an increase of between 10% and 80% (or more) by 2020 (S9.3.7).

■ Three out of four MA scenarios project that the global flux of nitrogen to coastal ecosystems will increase by a further 10–20% by 2030 (*medium certainty*). River nitrogen will not change in most industrial countries, while a 20–30% increase is projected for developing countries. This is a consequence of increasing nitrogen inputs to surface water associated with urbanization, sanitation, development of sewerage systems, and lagging wastewater treatment, as well as increasing food production and associated inputs of nitrogen fertilizer, animal manure, atmospheric nitrogen deposition, and biological nitrogen fixation in agricultural systems. Growing river nitrogen loads will lead to increased incidence of problems associated with eutrophication in coastal seas (S9.3.7).

APPENDIX B
EFFECTIVENESS OF ASSESSED RESPONSES

A response is considered to be effective when its assessment indicates that it has enhanced the particular ecosystem service (or, in the case of biodiversity, its conservation and sustainable use) and contributed to human well-being without significant harm to other ecosystem services or harmful impacts to other groups of people. A response is considered promising either if it does not have a long track record to assess but appears likely to succeed or if there are known means of modifying the response so that it can become effective. A response is considered problematic if its historical use indicates either that it has not met the goals related to service enhancement (or conservation and sustainable use of biodiversity) or that it has caused significant harm to other ecosystem services. Labeling a response as effective does not mean that the historical assessment has not identified problems or harmful trade-offs. Such trade-offs almost always exist, but they are not considered significant enough as to negate the effectiveness of the response. Similarly, labeling a response as problematic does not mean that there are no promising opportunities to reform the response in a way that can meet its policy goals without undue harm to ecosystem services.

The typology of responses presented in the Table in this Appendix is defined by the nature of the intervention, classified as follows: institutional and legal (I), economic and incentives (E), social and behavioral (S), technological (T),

and knowledge and cognitive (K). Note that the dominant class is given in the Table. The actors who make decisions to implement a response are governments at different levels, such as international (GI) (mainly through multilateral agreements or international conventions), national (GN), and local (GL); the business/industry sector (B); and civil society, which includes nongovernmental organizations (NGO), community-based and indigenous peoples organizations (C), and research institutions (R). The actors are not necessarily equally important.

The table includes responses assessed for a range of ecosystem services—food, fresh water, wood, nutrient management, flood and storm control, disease regulation, and cultural services. It also assesses responses for biodiversity conservation, integrated responses, and responses addressing one specific driver: climate change.

Response	Effectiveness			Notes	Type of Response	Required Actors
	Effective	Promising	Problematic			
Biodiversity conservation and sustainable use						
Protected areas		■		PAs are extremely important in biodiversity and ecosystem conservation programs, especially in sensitive environments that contain valuable biodiversity components. At global and regional scales, existing PAs are essential but not sufficient to conserve the full range of biodiversity. PAs need to be better located, designed, and managed to ensure representativeness and to deal with the impacts of human settlement within them, illegal harvesting, unsustainable tourism, invasive species, and climate change. They also need a landscape approach that includes protection outside of PAs. (R5)	I	GI GN GL NGO C R
Helping local people to capture biodiversity benefits			■	Providing incentives for biodiversity conservation in the form of benefits for local people (e.g., through products from single species or from ecotourism) has proved to be very difficult. Programs have been more successful when local communities have been in a position to make management decisions consistent with overall biodiversity conservation. "Win-win" opportunities for biodiversity conservation and benefits for local communities exist, but local communities can often achieve greater benefits from actions that lead to biodiversity loss. (R5)	E	GN GL B NGO C
Promoting better management of wild species as a conservation tool, including ex situ conservation		■		More effective management of individual species should enhance biodiversity conservation and sustainable use. "Habitat-based" approaches are critical, but they cannot replace "species-based" approaches. Zoos, botanical gardens, and other ex situ programs build support for conservation, support valuable research, and provide cultural benefits of biodiversity. (R5)	T S	GN S NGO R
Integrating biodiversity into regional planning		■		Integrated regional planning can provide a balance among land uses that promotes effective trade-offs among biodiversity, ecosystem services, and other needs of society. Great uncertainty remains as to what components of biodiversity persist under different management regimes, limiting the current effectiveness of this approach. (R5)	I	GN GL NGO
Encouraging private-sector involvement in biodiversity conservation		■		Many companies are preparing their own biodiversity action plans, managing their landholdings in ways that are more compatible with biodiversity conservation, supporting certification schemes that promote more sustainable use, and accepting their responsibility for addressing biodiversity issues. The business case that has been made for larger companies needs to be extended to other companies as well. (R5)	I	NG B NGO R
Including biodiversity issues in agriculture, forestry, and fisheries		■		More-diverse production systems can be as effective as low-diversity systems, or even more effective. And strategies based on more intensive production rather than on the expansion of the area allow for better conservation. (R5)	T	NG B
Designing governance approaches to support biodiversity			■	Decentralization of biodiversity management in many parts of the world has had variable results. The key to success is strong institutions at all levels, with secure tenure and authority at local levels essential to providing incentives for sustainable management. (R5)	I	GI GN GL R
Promoting international cooperation through multilateral environmental agreements		■		MEAs should serve as an effective means for international cooperation in the areas of biodiversity conservation and sustainable use. They cover the most pressing drivers and issues related to biodiversity loss. Better coordination among conventions would increase their usefulness. (R5,15)	I	GI GN
Environmental education and communication		■		Environmental education and communication programs have both informed and changed preferences for biodiversity conservation and have improved implementation of biodiversity responses. Providing the human and financial resources to undertake effective work in this area is a continuing barrier. (R5)	S	GN GL NGO C

Response	Effectiveness			Notes	Type of Response	Required Actors
	Effective	Promising	Problematic			
Food						
Globalization, trade, and domestic and international policies on food			▪	Government policies related to food production (price supports and various types of payments, or taxes) can have adverse economic, social, and environmental effects. (R6)	E	GI GN B
Knowledge and education	▪			Further research can make food production socially, economically, and environmentally sustainable. Public education should enable consumers to make informed choices about nutritious, safe, and affordable food. (R6)	S K	GN GL NGO C
Technological responses, including biotechnology, precision agriculture, and organic farming		▪		New agricultural sciences and effective natural resource management could support a new agricultural revolution to meet worldwide food needs. This would help environmental, economic, and social sustainability. (R6)	T	GN B R
Water management		▪		Emerging water pricing schemes and water markets indicate that water pricing can be a means for efficient allocation and responsible use. (R6)	E	GN GL B NGO
Fisheries management		▪		Strict regulation of marine fisheries both regarding the establishment and implementation of quotas and steps to address unreported and unregulated harvest. Individual transferable quotas also show promise for coldwater, single-species fisheries but they are unlikely to be useful in multispecies tropical fisheries. Given the potential detrimental environmental impacts of aquaculture, appropriate regulatory mechanisms need to supplement existing polices. (R6)	I E	GN GL B NGO
Livestock management			▪	Livestock polices need to be reoriented in view of problems concerning overgrazing and dryland degradation, rangeland fragmentation and loss of wildlife habitat, dust formation, bush encroachment, deforestation, nutrient overload through disposal of manure, and greenhouse gas emissions. Policies also need to focus on human health issues related to diseases such as bird flu and BSE. (R6)	T	GN B
Recognition of gender issues		▪		Response policies need to be gender-sensitive and designed to empower women and ensure access to and control of resources necessary for food security. This needs to be based on a systematic analysis of gender dynamics and explicit consideration of relationships between gender and food and water security. (R6)	S	GN NGO C
Fresh water						
Determining ecosystem water requirements		▪		In order to balance competing demands, it is critical that society explicitly agrees on ecosystem water requirements (environmental flows). (R7)	I T	GN GL NGO R
Rights to freshwater services and responsibilities for their provision		▪		Both public and private ownership systems of fresh water and of the land resources associated with its provision have largely failed to create incentives for provision of services. As a result, upland communities have generally been excluded from access to benefits, particularly when they lack tenure security, and have resisted regulations regarded as unfair. Effective property rights systems with clear and transparent rules can increase stakeholders' confidence that they will have access to the benefits of freshwater services and, therefore, their willingness to pay for them. (R7)	I	GN B C

(continued on page 126)

Response	Effectiveness			Notes	Type of Response	Required Actors
	Effective	Promising	Problematic			
Fresh water *(continued)*						
Increasing the effectiveness of public participation in decision-making		●		Degradation of fresh water and other ecosystem services has a disproportionate impact on those excluded from participation in decision-making. Key steps for improving participatory processes are to increase the transparency of information, improve the representation of marginalized stakeholders, engage them in the establishment of policy objectives and priorities for the allocation of freshwater services, and create space for deliberation and learning that accommodates multiple perspectives. (R7)	I	GN GL NGO C R
River basin organizations		●		RBOs can play an important role in facilitating cooperation and reducing transaction costs of large-scale responses. RBOs are constrained or enabled primarily by the degree of stakeholder participation, their agreement on objectives and management plans, and their cooperation on implementation. (R7)	I	GI GN NGO
Regulatory responses	●			Regulatory approaches based on market-based incentives (e.g., damages for exceeding pollution standards) are suitable for point-source pollutants. Regulatory approaches that simply outlaw particular types of behavior can be unwieldy and burdensome and may fail to provide incentives for protecting freshwater services. (R7)	I	GN GL
Water markets		●		Economic incentives can potentially unlock significant supply- and demand-side efficiencies while providing cost-effective reallocation between old (largely irrigation) and new (largely municipal and instream) uses. (R7)	E	GI GN B
Payments for watershed services		●		Payments for ecosystem services provided by watersheds have narrowly focused on the role of forests in the hydrological regime. They should be based on the entire flow regime, including consideration of the relative values of other land cover and land uses, such as wetlands, riparian areas, steep slopes, roads, and management practices. Key challenges for payment schemes are capacity building for place-based monitoring and assessment, identifying services in the context of the entire flow regime, considering trade-offs and conflicts among multiple uses, and making uncertainty explicit. (R7)	E	GN B C
Partnerships and financing		●		There is a clear mismatch between the high social value of freshwater services and the resources allocated to manage water. Insufficient funding for water infrastructure is one manifestation of this. Focusing only on large-scale privatization to improve efficiency and cost-recovery has proved a double-edged strategy—price hikes or control over resources have created controversy and, in some cases, failure and withdrawal. Development of water infrastructure and technologies must observe best practices to avoid problems and inequities. The reexamination and retrofitting/refurbishment of existing infrastructure is the best option in the short and medium term. (R7)	I E	GI GN B NGO C
Large dams			●	The impact of large dams on freshwater ecosystems is widely recognized as being more negative than positive. In addition, the benefits of their construction have rarely been shared equitably—the poor and vulnerable and future generations often fail to receive the social and economic benefits from dams. Preconstruction studies typically are overly optimistic about the benefits of projects and underestimate costs. (R7)	T	GN
Wetland restoration			●	Although wetland restoration is a promising management approach, there are significant challenges in determining what set of management interventions will produce a desired combination of wetland structure and function. It is unlikely that created wetlands can structurally and functionally replace natural wetlands. (R7)	T	GN GL NGO B

Response	Effectiveness			Notes	Type of Response	Required Actors
	Effective	Promising	Problematic			
Wood, fuelwood, and non-wood forest products						
International forest policy processes and development assistance				International forest policy processes have made some gains within the forest sector. Attention should be paid to integration of agreed forest management practices in financial institutions, trade rules, global environment programs, and global security decision-making. (R8)	I	GI GN B
Trade liberalization				Forest product trade tends to concentrate decision-making power on (and benefits from) forest management rather than spreading it to include poorer and less powerful players. It "magnifies" the effect of governance, making good governance better and bad governance worse. Trade liberalization can stimulate a "virtuous cycle" if the regulatory framework is robust and externalities are addressed. (R8)	E	GI GN
National forest governance initiatives and national forest programs				Forest governance initiatives and country-led national forest programs show promise for integrating ecosystem health and human well-being where they are negotiated by stakeholders and strategically focused. (R8)	I	GN GL
Direct management of forests by indigenous peoples				Indigenous control of traditional homelands is often presented as having environmental benefits, although the main justification continues to be based on human and cultural rights. Little systematic data exist, but preliminary findings on vegetation cover and forest fragmentation from the Brazilian Amazon suggest that an indigenous control area can be at least as effective as a strict-use protected area. (R8)	I	GL C
Collaborative forest management and local movements for access and use of forest products				Government-community collaborative forest management can be highly beneficial but has had mixed results. Programs have generated improved resource management access of the rural poor to forest resources but have fallen short in their potential to benefit the poor. Local responses to problems of access and use of forest products have proliferated in recent years. They are collectively more significant than efforts led by governments or international processes but require their support to spread. (R8)	I	GN GL B NGO C
Small-scale private and public-private ownership and management of forests				Where information, tenure, and capacity are strong, small private ownership of forests can deliver more local economic benefits and better forest management than ownership by larger corporate bodies. (R8)	I	GL B C
Company-community forestry partnerships				Company-community partnerships can be better than solely corporate forestry, or than solely community or small-scale farm forestry, in delivering benefits to the partners and the public at large. (R8)	I	GL B C
Public and consumer action				Public and consumer action has resulted in important forest and trade policy initiatives and improved practices in large forest corporations. This has had an impact in "timber-consuming countries" and in international institutions. The operating standards of some large corporations and institutions, as well as of those whose non-forest activities have an impact on forests, have been improved. (R8)	S	NGO B C
Third-party voluntary forest certification				Forest certification has become widespread; however, most certified forests are in the North, managed by large companies, and exporting to northern retailers. The early proporents of certification hoped it would be an effective response to tropical deforestation. (R8)	I E	B
Wood technology and biotechnology				Wood technology responses have focused on industrial plantation species with properties suited for manufactured products. (R8)	T	NG R B

(continued on page 128)

Response	Effectiveness			Notes	Type of Response	Required Actors
	Effective	Promising	Problematic			
Wood, fuelwood, and non-wood forest products *(continued)*						
Commercialization of non-wood forest products				Commercialization of NWFPs has had modest impacts on local livelihoods and has not always created incentives for conservation. An increased value of NWFPs is not always an incentive for conservation and can have the opposite effect. Incentives for sustainable management of NWFPs should be reconsidered, including exploration of joint-production of timber and NWFPs. (R8)	E	NGO B R
Natural forest management in the tropics				To be economic, sustainable natural forest management in the tropics must focus on a range of forest goods and services, not just timber. The "best practices" of global corporations should be assessed, exploring at the same time "what works" in traditional forest management and the work of local (small) enterprises. Considerable interest has developed in the application of reduced-impact logging, especially in tropical forests, which lowers environmental impacts and can also be more efficient and cost-effective. (R8)	T	GI GN GL B NGO C
Forest plantation management				Farm woodlots and large-scale plantations are increasingly being established in a response to growing wood demand and declining natural forest areas. Without adequate planning and management, forest plantations can be established in the wrong sites, with the wrong species and provenances. In degraded lands, afforestation may deliver economic, environmental, and social benefits to communities and help reduce poverty and enhancing food security. (R8)	T	GN GL B NGO R
Fuelwood management				Fuelwood remains one of the main products of the forest sector in the South. If technology development continues, industrial-scale forest product fuels could become a major sustainable energy source. (R8)	T	GL B C
Afforestation and reforestation for carbon management				Although many early initiatives were based on forest conservation or management, afforestation activities now predominate, perhaps reflecting the international decision in 2001 to allow only afforestation and reforestation activities into the Clean Development Mechanism for the first commitment period. (R8)	T E	GI GN B
Nutrient cycling						
Regulations				Mandatory policies, including regulatory control and tax or fee systems, place the costs and burden of pollution control on the polluter. Technology-based standards are easy to implement but may discourage innovation and are generally not seen as cost-effective. (R9)	I	GI GN
Market-based instruments				Market-based instruments, such as financial incentives, subsidies, and taxes, hold potential for better nutrient management but may not be relevant in all countries and circumstances. Relatively little is known empirically about the impact of these instruments on technological change. (R9)	E	GN B R
Hybrid approaches				Combinations of regulatory, incentive, and market-based mechanisms are possible for both national and watershed-based approaches and may be the most cost-effective and politically acceptable. (R9)	I E	GI GN GL NGO C, R
Flood and storm regulation						
Physical structures				Historically, emphasis was on physical structures and measures over natural environment and social institutions. This choice often creates a false sense of security, encouraging people to accept high risks. Evidence indicates that more emphasis needs to be given to the natural environment and nonstructural measures. (R11)	T	GN B

Response	Effectiveness			Notes	Type of Response	Required Actors
	Effective	Promising	Problematic			
Flood and storm regulation (continued)						
Use of natural environment				Flood and storm impacts can be lessened through maintenance and management of vegetation and through natural or humanmade geomorphological features (natural river channels, dune systems, terrace farming). (R11)	T	GN GL NGO C
Information, institutions, and education				These approaches, which emphasize disaster preparedness, disaster management, flood and storm forecasting, early warning, and evacuation, are vital for reducing losses. (R11)	S I	GN GL B C
Financial services				These responses emphasize insurance, disaster relief, and aid. Both social programs and private insurance are important coping mechanisms for flood disaster recovery. They can, however, inadvertently contribute to community vulnerability by encouraging development within floodplains or by creating cultures of entitlement. (R11)	E	GN B
Land use planning				Land use planning is a process of determining the most desirable type of land use. It can help mitigate disasters and reduce risks by avoiding development in hazard-prone areas. (R11)	I	GN
Disease regulation						
Integrated vector management				Reducing the transmission of infectious diseases often has effects on other ecosystems services. IVM enables a coordinated response to health and the environment. IVM uses targeted interventions to remove or control vector-breeding sites, disrupt vector lifecycles, and minimize vector-human contact, while minimizing effects on other ecosystem services. IVM is most effective when integrated with socioeconomic development. (R12)	I	GN NGO
Environmental management or modification to reduce vector and reservoir host abundance				Environmental management can be highly cost-effective and entail very low environmental impacts. (R12)	I	GN B C R
Biological control or natural predators				Biological interventions can be highly cost-effective and entail very low environmental impacts. Biological control may be effective if breeding sites are well known and limited in number but less feasible where they are numerous. (R12)	T	GN B R
Chemical control				Insecticides remain an important tool and their selective use is likely to continue within IVM. However, there are concerns regarding the impacts of insecticides, especially persistent organic pollutants, on the environment and on human populations, particularly insecticide sprayers. (R12)	T	GN B R
Human settlement patterns				The most basic management of human-vector contact is through improvements in the placement and construction of housing. (R12)	T	GN NGO C
Health awareness and behavior				Social and behavioral responses can help control vector-borne disease while also improving other ecosystem services. (R12)	S	C
Genetic modification of vector species to limit disease transmission				New "cutting-edge" interventions, such as transgenic techniques, could be available within the next 5–10 years. However, consensus is lacking in the scientific community on the technical feasibility and public acceptability of such an approach. (R12)	T	GN B NGO R

(*continued on page 130*)

Response	Effectiveness			Notes	Type of Response	Required Actors
	Effective	Promising	Problematic			
Cultural services						
Awareness of the global environment and linking local and global institutions	■			Awareness of the planet working as a system has led to an integrated approach to ecosystems. This process has emphasized the "human environment" concept and the discussion of environmental problems at a global scale. Local organizations also take advantage of emerging global institutions and conventions to bring their case to wider political arenas. (R14)	S I	GI GN GL
From restoring landscapes to valuing cultural landscapes		■		Landscapes are subject to and influenced by cultural perceptions and political and economic interests. This influences decisions on landscape conservation. (R14)	S K	GL NGO C
Recognizing sacred areas		■		While linking sacred areas and conservation is not new, there has been an increase in translating "the sacred" into legislation or legal institutions granting land rights. This requires extensive knowledge about the link between the sacred, nature, and society in a specific locale. (R14)	S	GL NGO C
International agreements and conservation of biological and agropastoral diversity		■		Increased exploitation and awareness concerning the disappearance of local resources and knowledge has highlighted the need to protect local and indigenous knowledge. Some countries have adopted specific laws, policies, and administrative arrangements emphasizing the concept of prior informed consent of knowledge-holders. (R14)	I	GI GN
Integrating local and indigenous knowledge			■	Local and indigenous knowledge evolves in specific contexts, and good care should be taken to not de-contextualize it. Conventional "best-practices" methods focusing on content may not be appropriate to deal with local or indigenous knowledge. (R14)	K I	GN B NGO
Compensating for knowledge			■	Compensation for the use of local and indigenous knowledge by third parties is an important yet complicated response. The popular idea that local and indigenous knowledge can be promoted by strengthening "traditional" authorities may not be valid in many cases. (R14)	E K	GN B C
Property right changes			■	Communities benefit from control over natural resources but traditional leadership may not always be the solution. Local government institutions that are democratically elected and have real authority over resources may be in some cases a better option. There is a tendency to shift responsibilities back and forth between "traditional" authorities and local government bodies, without giving any of them real decision-making powers. (R14)	I	GN GL C
Certification programs		■		Certification programs are a promising response, but many communities do not have access to these programs or are not aware of their existence. In addition, the financial costs involved reduce the chances for local communities to participate independently. (R14)	I S	GI GN B
Fair trade		■		Fair trade is a movement initiated to help disadvantaged or politically marginalized communities by paying better prices and providing better trading conditions, along with raising consumers' awareness of their potential role as buyers. Fair trade overlaps in some cases with initiatives focusing on the environmental performance of trade. (R14)	E S	GI GN GL NGO C
Ecotourism and cultural tourism		■		Ecotourism can provide economic alternatives to converting ecosystems, however it can generate conflicts in resource use and the aesthetics of certain ecosystems. Different ecosystems are subjected to different types and scales of impact from tourism infrastructure. Furthermore, some ecosystems are easier to market to tourists than others. The market value of ecosystems may vary according to public perceptions of nature. Freezing of landscapes, conversion of landscapes, dispossession, and removing of human influences may result, depending on views of what ecotourism should represent. Yet when conservation receives no budgetary subsidy, tourism can provide revenues for conservation. (R14)	E	GL B C

Response	Effectiveness			Notes	Type of Response	Required Actors
	Effective	Promising	Problematic			
Integrated responses						
International environmental governance				Environmental policy integration at the international level is almost exclusively dependent on governments' commitment to binding compromises on given issues. Major challenges include reform of the international environmental governance structure and coherence between international trade and environment mechanisms. (R15)	I E K T B	GI GN
National action plans and strategies aiming to integrate environmental issues into national policies				Examples include National Conservation Strategies, National Environmental Action Plans, and National Strategies for Sustainable Development. Success depends on enabling conditions such as ownership by governments and civil society and broad participation, both across sectors within the government and with the private sector as well as at sub-national and local scales. National integrated responses may be a good starting point for cross-departmental linkages in governments. (R15)	I E K T B	GN GL B NGO C
Sub-national and local integrated approaches				Many integrated responses are implemented at the sub-national level, and examples include sustainable forest management, integrated coastal zone management, integrated conservation and development programs, and integrated river basin management. Results so far have been varied, and a major constraint experienced by sub-national and multiscale responses is the lack of implementation capacity. (R15)	I E K T B	GN GL NGO C
Climate change						
U.N. Framework Convention on Climate Change and Kyoto Protocol				The ultimate goal of UNFCCC is stabilization of greenhouse gas concentrations in the atmosphere at a level that would prevent dangerous anthropogenic interference with the climate system. The Kyoto Protocol contains binding limits on greenhouse gas emissions on industrial countries that agreed to reduce their emissions by an average of about 5% between 2008 and 2012 relative to the levels emitted in 1990. (R13)	I	GI GN
Reductions in net greenhouse gas emissions				Significant reductions in net greenhouse gas emissions are technically feasible, in many cases at little or no cost to society. (R13)	T	GN B C
Land use and land cover change				Afforestation, reforestation, improved management of forests, croplands, and range-lands, and agroforestry provide opportunities to increase carbon uptake, and slowing deforestation reduces emissions. (R13)	T	GN GL B NGO C
Market mechanisms and incentives				The Kyoto Protocol mechanisms, in combination with national and regional ones, can reduce the costs of mitigation for industrial countries. In addition, countries can reduce net costs of emissions abatement by taxing emissions (or auctioning permits) and using the revenues to cut distortion taxes on labor and capital. In the near term, project-based trading can facilitate the transfer of climate-friendly technologies to developing countries. (R13)	E	GI GN B
Adaptation				Some climate change is inevitable, and ecosystems and human societies will need to adapt to new conditions. Human populations will face the risk of damage from climate change, some of which may be countered with current coping systems; others may need radically new behaviors. Climate change needs to be factored into current development plans. (R13)	I	GN GL NGO C R

APPENDIX C
AUTHORS, COORDINATORS, AND REVIEW EDITORS

Core Writing Team

WALTER V. REID, Millennium Ecosystem Assessment, Malaysia and United States

HAROLD A. MOONEY, Stanford University, United States

ANGELA CROPPER, The Cropper Foundation, Trinidad and Tobago

DORIS CAPISTRANO, Center for International Forestry Research, Indonesia

STEPHEN R. CARPENTER, University of Wisconsin - Madison, United States

KANCHAN CHOPRA, Institute of Economic Growth, India

PARTHA DASGUPTA, University of Cambridge, United Kingdom

THOMAS DIETZ, Michigan State University, United States

ANANTHA KUMAR DURAIAPPAH, International Institute for Sustainable Development, Canada

RASHID HASSAN, University of Pretoria, South Africa

ROGER KASPERSON, Clark University, United States

RIK LEEMANS, Wageningen University, Netherlands

ROBERT M. MAY, University of Oxford, United Kingdom

TONY (A.J.) MCMICHAEL, Australian National University, Australia

PRABHU PINGALI, Food and Agriculture Organization of the United Nations, Italy

CRISTIÁN SAMPER, National Museum of Natural History, United States

ROBERT SCHOLES, Council for Science and Industrial Research, South Africa

ROBERT T. WATSON, The World Bank, United States

A.H. ZAKRI, United Nations University, Japan

ZHAO SHIDONG, Chinese Academy of Sciences, China

NEVILLE J. ASH, UNEP-World Conservation Monitoring Centre, United Kingdom

ELENA BENNETT, University of Wisconsin - Madison, United States

PUSHPAM KUMAR, Institute of Economic Growth, India

MARCUS LEE, WorldFish Center, Malaysia

CIARA RAUDSEPP-HEARNE, Millennium Ecosystem Assessment, Malaysia

HENK SIMONS, National Institute of Public Health and the Environment, Netherlands

JILLIAN THONELL, UNEP-World Conservation Monitoring Centre, United Kingdom

MONIKA B. ZUREK, Food and Agriculture Organization of the United Nations, Italy

Millennium Ecosystem Assessment Coordinating Lead Authors, Conceptual Framework Lead Authors, and Sub-global Assessment Coordinators

ADEL FARID ABDEL-KADER, United Nations Environment Programme, Bahrain

NIMBE ADEDIPE, National Universities Commission, Nigeria

ZAFAR ADEEL, United Nations University - International Network on Water, Environment and Health, Canada

JOHN B.R. AGARD, University of the West Indies, Trinidad and Tobago

TUNDI AGARDY, Sound Seas, United States

HEIDI ALBERS, Oregon State University, United States

JOSEPH ALCAMO, University of Kassel, Germany

JACQUELINE ALDER, University Of British Columbia, Canada

MOURAD AMIL, Ministere de l'Amenagement du Territoire, de l'Eau et de l'Environnement, Morocco

ALEJANDRO ARGUMEDO, Asociacion Kechua-Aymara ANDES, Peru

DOLORS ARMENTERAS, Instituto de Investigacion de Recursos Biológicos Alexander von Humboldt, Colombia

NEVILLE J. ASH, UNEP-World Conservation Monitoring Centre, United Kingdom

BRUCE AYLWARD, Deschutes Resources Conservancy, United States

SURESH CHANDRA BABU, International Food Policy Research Institute, India

JAYANTA BANDYOPADHYAY, Indian Institute of Management, India

CHARLES VICTOR BARBER, IUCN – World Conservation Union, United States

STEPHEN BASS, Department for International Development, United Kingdom

ALLAN BATCHELOR, B&M Environmental Services (Pty) Ltd, South Africa

T. DOUGLAS BEARD, JR., United States Geological Survey, United States

ANDREW BEATTIE, Macquarie University, Australia

JUAN CARLOS BELAUSTEGUIGOITIA, Global International Waters Assessment, Sweden

ELENA BENNETT, University of Wisconsin - Madison, United States

D.K. BHATTACHARYA, University of Delhi, India

HERNÁN BLANCO, Recursos e Investigación para el Desarrollo Sustentable, Chile

JORGE E. BOTERO, Centro Nacional de Investigaciones de Café, Colombia

LELYS BRAVO DE GUENNI, Universidad Simón Bolívar, Venezuela

EDUARDO BRONDIZIO, Indiana University, United States

VICTOR BROVKIN, Potsdam Institute for Climate Impact Research, Germany

KATRINA BROWN, University of East Anglia, United Kingdom

COLIN D. BUTLER, Australian National University, Australia

J. BAIRD CALLICOTT, University of North Texas, United States

ESTHER CAMAC-RAMIREZ, Association Ixä Ca Vaá for Indigenious Development and Information, Costa Rica

DIARMID CAMPBELL-LENDRUM, World Health Oganization, Switzerland

DORIS CAPISTRANO, Center for International Forestry Research, Indonesia

FABRICIO WILLIAM CARBONELL TORRES, Association Ixä Ca Vaá for Indigenious Development and Information, Costa Rica

STEPHEN R. CARPENTER, University of Wisconsin - Madison, United States

KENNETH G. CASSMAN, University of Nebraska - Lincoln, United States

JUAN CARLOS CASTILLA, Center for Advance Studies in Ecology and Biodiversity, Chile

ROBERT CHAMBERS, Institute of Development Studies - Sussex, United Kingdom

W. BRADNEE CHAMBERS, United Nations University, Japan

F. STUART CHAPIN, III, University of Alaska - Fairbanks, United States

KANCHAN CHOPRA, Institute of Economic Growth, India

FLAVIO COMIM, University of Cambridge, United Kingdom and Federal University of Rio Grande do Sul, Brazil

ULISSES E.C. CONFALONIERI, National School of Public Health, Brazil

STEVE CORK, Land and Water Australia, Australia

CARLOS CORVALAN, World Health Organization, Switzerland

WOLFGANG CRAMER, Potsdam Institute for Climate Impact Research, Germany

ANGELA CROPPER, The Cropper Foundation, Trinidad and Tobago

GRAEME CUMMING, University of Florida, United States

OWEN CYLKE, World Wildlife Fund, United States

REBECCA D'CRUZ, Aonyx Environmental, Malaysia

GRETCHEN C. DAILY, Stanford University, United States

PARTHA DASGUPTA, University of Cambridge, United Kingdom

RUDOLF S. DE GROOT, Wageningen University, Netherlands

RUTH S. DEFRIES, University of Maryland, United States

SANDRA DIAZ, Universidad Nacional de Córdoba, Argentina

THOMAS DIETZ, Michigan State University, United States

RICHARD DUGDALE, San Francisco State University, United States

ANANTHA KUMAR DURAIAPPAH, International Institute for Sustainable Development, Canada

SIMEON EHUI, The World Bank, United States

POLLY ERICKSEN, Columbia University Earth Institute, United States

CHRISTO FABRICIUS, Rhodes University, South Africa

DAN FAITH, Australian Museum, Australia

JOSEPH FARGIONE, University of New Mexico, United States

COLIN FILER, Australian National University, Australia

C. MAX FINLAYSON, Environmental Research Institute of the Supervising Scientist, Australia

DANA R. FISHER, Columbia University, United States

CARL FOLKE, Stockholm University, Sweden

MIGUEL FORTES, Intergovernmental Oceanographic Commission Regional Secretariat for the Western Pacific, Thailand

MADHAV GADGIL, Indian Institute of Science, India

HABIBA GITAY, Australian National University, Australia

YOGESH GOKHALE, Indian Institute of Science, India

THOMAS HAHN, Stockholm University, Sweden

SIMON HALES, Wellington School of Medicine & Health Sciences, New Zealand

KIRK HAMILTON, The World Bank, United States

RASHID HASSAN, University of Pretoria, South Africa

HE DAMING, Yunnan University, China

KENNETH R. HINGA, United States Department of Agriculture, United States

ANKILA J. HIREMATH, Ashoka Trust for Research in Ecology and the Environment, India

JOANNA HOUSE, Max Planck Institute for Biogeochemistry, Germany

ROBERT W. HOWARTH, Cornell University, United States

TARIQ ISMAIL, Saudi Arabia

ANTHONY JANETOS, The H. John Heinz III Center for Science, Economics, and the Environment, United States

PETER KAREIVA, The Nature Conservancy, United States

ROGER KASPERSON, Clark University, United States

KISHAN KHODAY, United Nations Development Programme, Indonesia

CHRISTIAN KOERNER, University of Basel, Switzerland

KASPER KOK, Wageningen University, Netherlands

PUSHPAM KUMAR, Institute of Economic Growth, India

ERIC F. LAMBIN, Universite Catholique de Louvain, Belgium

PAULO LANA, Universidade Federal do Paraná, Brazil

RODEL D. LASCO, World Agroforestry Centre, Philippines

PATRICK LAVELLE, University of Paris VI/ IRD, France

LOUIS LEBEL, Chiang Mai University, Thailand

MARCUS LEE, WorldFish Center, Malaysia

RIK LEEMANS, Wageningen University, Netherlands

CHRISTIAN LÉVÊQUE, Institut de Recherches pour le développement, France

MARC LEVY, Columbia University, United States

LIU JIAN, Chinese Academy of Sciences, China

LIU JIYUAN, Chinese Academy of Sciences, China

MA SHIMING, Chinese Academy of Agricultural Sciences, China

GEORGINA MACE, Zoological Society of London, United Kingdom

JENS MACKENSEN, United Nations Environment Programme, Kenya

MAI TRONG THONG, Vietnamese Academy of Science and Technology, Vietnam

BEN MALAYANG III, Philippine Sustainable Development Network and University of the Philippines Los Baños, Philippines

JEAN-PAUL MALINGREAU, Joint Research Centre of the European Commission, Belgium

ANATOLY MANDYCH, Russian Academy of Sciences, Russian Federation

PETER JOHN MARCOTULLIO, United Nations University, Japan

EDUARDO MARONE, Centro de Estudos do Mar, Brazil

HILLARY M. MASUNDIRE, University of Botswana, Botswana

ROBERT M. MAY, University of Oxford, United Kingdom

JAMES MAYERS, International Institute for Environment and Development, United Kingdom

ALEX F. MCCALLA, University of California - Davis, United States

JACQUELINE MCGLADE, European Environment Agency, Denmark

GORDON MCGRANAHAN, International Institute for Environment and Development, United Kingdom

TONY (A.J.) MCMICHAEL, Australian National University, Australia

JEFFREY A. MCNEELY, IUCN-The World Conservation Union, Switzerland

MONIRUL Q. MIRZA, University of Toronto, Canada

BEDRICH MOLDAN, Charles University, Czech Republic

DAVID MOLYNEUX, Liverpool School of Tropical Medicine, United Kingdom

HAROLD A. MOONEY, Stanford University, United States

SANZHAR MUSTAFIN, Regional Environmental Centre for Central Asia, Kazakhstan

CONSTANCIA MUSVOTO, University of Zimbabwe, Zimbabwe

SHAHID NAEEM, Columbia University, United States

NEBOJŠA NAKIĆENOVIĆ, International Institute for Applied Systems Analysis, Austria

GERALD C. NELSON, University of Illinois - Urbana-Champaign, United States

NIU WEN-YUAN, Chinese Academy of Sciences, China

IAN NOBLE, The World Bank, United States

SIGNE NYBØ, Norwegian Institute for Nature Research, Norway

MASAHIKO OHSAWA, University of Tokyo, Japan

WILLIS OLUOCH-KOSURA, University of Nairobi, Kenya

OUYANG ZHIYUN, Chinese Academy of Sciences, China

STEFANO PAGIOLA, The World Bank, United States

CHERYL A. PALM, Columbia University, United States

JYOTI K. PARIKH, Integrated Research and Action for Development, India

ANAND PATWARDHAN, Indian Institute of Technology-Bombay, India

ANKUR PATWARDHAN, Research & Action in Natural Wealth Administration, India

JONATHAN PATZ, University of Wisconsin - Madison, United States

DANIEL PAULY, University of British Columbia, Canada

STEVE PERCY, United States

HENRIQUE MIGUEL PEREIRA, University of Lisbon, Portugal

REIDAR PERSSON, Swedish University of Agricultural Sciences, Sweden

GARRY D. PETERSON, McGill University, Canada

GERHARD PETSCHEL-HELD, Potsdam Institute for Climate Impact Research, Germany

INA BINARI PRANOTO, Ministry of Environment, Indonesia

ROBERT PRESCOTT-ALLEN, Coast Information Team, Canada

RUDY RABBINGE, Wageningen University, Netherlands

KILAPARTI RAMAKRISHNA, Woods Hole Research Center, United States

P. S. RAMAKRISHNAN, Jawaharlal Nehru University, India

PAUL RASKIN, Tellus Institute, United States

CIARA RAUDSEPP-HEARNE, Millennium Ecosystem Assessment, Malaysia

WALTER V. REID, Millennium Ecosystem Assessment, Malaysia and United States

CARMEN REVENGA, The Nature Conservancy, United States

BELINDA REYERS, Council for Science and Industrial Research, South Africa

TAYLOR H. RICKETTS, World Wildlife Fund, United States

JANET RILEY, Rothamsted Research, United Kingdom

CLAUDIA RINGLER, International Food Policy Research Institute, United States

JON PAUL RODRIGUEZ, Instituto Venezolano de Investigaciones, United States

JEFFREY M. ROMM, University of California Berkeley, United States

SERGIO ROSENDO, University of East Anglia, United Kingdom

URIEL N. SAFRIEL, Hebrew University of Jerusalem, Israel

OSVALDO E. SALA, Brown University, United States

CRISTIÁN SAMPER, National Museum of Natural History, United States

NEIL SAMPSON, The Sampson Group, Inc., United States

ROBERT SCHOLES, Council for Science and Industrial Research, South Africa

MAHENDRA SHAH, International Institute for Applied System Analysis, Austria

ALEXANDER SHESTAKOV, World Wildlife Fund Russian Programme, Russian Federation

ANATOLY SHVIDENKO, Institute for Applied Systems Analysis, Austria

HENK SIMONS, National Institute of Public Health and the Environment, Netherlands

DAVID SIMPSON, United States Environmental Protection Agency, United States

NIGEL SIZER, The Nature Conservancy, Indonesia

MARJA SPIERENBURG, Vrije Universiteit Amsterdam, Netherlands

BIBHAB TALUKDAR, Ashoka Trust for Research in Ecology and the Environment, India

MOHAMED TAWFIC AHMED, Suez Canal University, Egypt

PONGMANEE THONGBAI, Thailand Institute of Scientific and Technological Research, Thailand

DAVID TILMAN, University of Minnesota, United States

THOMAS P. TOMICH, World Agroforestry Centre, Kenya

FERENC L. TOTH, International Atomic Energy Agency, Austria

JANE K. TURPIE, University of Cape Town, South Africa

ALBERT S. VAN JAARSVELD, Stellenbosch University, South Africa

DETLEF VAN VUUREN, National Institute for Public Health and the Environment, Netherlands

JOELI VEITAYAKI, University of the South Pacific, Fiji

SANDRA J. VELARDE, World Agroforestry Centre, Kenya

RODRIGO A. BRAGA MORAES VICTOR, São Paulo City Green Belt Biosphere Reserve - Forest Institute, Brazil

ERNESTO F. VIGLIZZO, National Institute for Agricultural Technology, Argentina

BHASKAR VIRA, University of Cambridge, United Kingdom

CHARLES J. VÖRÖSMARTY, University of New Hampshire, United States

DIANA HARRISON WALL, Colorado State University, United States

MERRILYN WASSON, Australian National University, Australia

MASATAKA WATANABE, National Institute for Environmental Studies, Japan

ROBERT T. WATSON, The World Bank, United States

THOMAS J. WILBANKS, Oak Ridge National Laboratory, United States

MERYL WILLIAMS, Consultative Group on International Agricultural Research, Malaysia

POH POH WONG, National University of Singapore, Singapore

STANLEY WOOD, International Food Policy Research Institute, United States

ELLEN WOODLEY, Terralingua, Canada

ALISTAIR WOODWARD, University of Auckland, New Zealand

ANASTASIOS XEPAPADEAS, University of Crete, Greece

GARY YOHE, Wesleyan University, United States

YUE TIANXIANG, Chinese Academy of Sciences, China

MARIA FERNANDA ZERMOGLIO, University of California - Davis, United States

ZHAO SHIDONG, Chinese Academy of Sciences, China

MONIKA B. ZUREK, Food and Agriculture Organization of the United Nations, Italy

APPENDIX D
ABBREVIATIONS, ACRONYMS, AND FIGURE SOURCES

Abbreviations and Acronyms

BSE – bovine spongiform encephalopathy

CBD – Convention on Biological Diversity

DALY – disability-adjusted life year

FAO – Food and Agriculture Organization (United Nations)

GDP – gross domestic product

GHS – greenhouse gases

GNI – gross national income

GNP – gross national product

IPCC – Intergovernmental Panel on Climate Change

IUCN – World Conservation Union

IVM – integrated vector management

MA – Millennium Ecosystem Assessment

MEA – multilateral environmental agreement

MDG – Millennium Development Goal

NGO – nongovernmental organization

NPP – net primary productivity

NWFP – non-wood forest product

OECD – Organisation for Economic Co-operation and Development

PA – protected area

RBO – river basin organization

SARS – severe acute respiratory syndrome

SCOPE – Scientific Committee on Problems of the Environment

UNCCD – United Nations Convention to Combat Desertification

UNEP – United Nations Environment Programme

UNFCCC – United Nations Framework Convention on Climate Change

WWF – World Wide Fund for Nature

Chemical Symbols, Compounds, Units of Measurement

CH_4 – methane

CO – carbon monoxide

CO_2 – carbon dioxide

GtC-eq – gigatons of carbon equivalent

N – nitrogen

N_2O – nitrous oxide

NOx – nitrogen oxides

ppmv – parts per million by volume

SO_2 – sulfur dioxide

teragram – 10^{12} grams

Figure Sources

Most Figures used in this report were redrawn from Figures included in the technical assessment reports in the chapters referenced in the Figure captions. Preparation of several Figures involved additional information as follows:

Figure 11 (and Figure 3.4)
> The source Figure from CF Box 2.4 was updated to 2003/04 with data from Northern Cod (2J+3KL) Stock Status Update, Fisheries and Oceans Canada, March 2004.

Figure 14 (and Figure 1.5)
> The source Figure (R9 Fig 9.1) was modified to include the addition of projected human inputs in 2050 based on data included in the original source for R9 Fig 9.1: Galloway, J.P., et al., 2004, *Biogeochemistry* 70: 153–226.

Figure 1.6
> The source Figure (R9 Fig 9.2) was modified to include two additional deposition maps for 1860 and 2050 that had been included in the original source for R9 Fig 9.2: Galloway, J.P., et al., 2004, *Biogeochemistry* 70: 153–226.

Figure 1.7
> This Figure was developed from two Figures included in articles cited in C11.3.1: Ruiz et al., 2000, *Annual Review of Ecology and Systematics* 31: 481-531 (Fig 1c); Ribera Siguan 2003, in G.M. Ruiz and J.T. Carlton eds., *Invasive Species: Vectors and Management Strategies*, Island Press, Washington D.C. (Fig 8.5).

Figures B and C in Box 3.1 - Linkages between Ecosystem Services and Human Well-being
> The source Figures (C7 Fig 7.13 and 7.14) are based on World Health Organization and United Nations Children's Fund, 2000: *Global Water Supply and Sanitation Assessment 2000 Report*, World Health Organization, Geneva, updated for 2002 using the WHO online database.

Figure 3.1
> This Figure was developed from the database cited in C5.2.6 using World Bank figures for "adjusted net savings" for 2001, downloaded from lnweb18.worldbank.org/ESSD/envext.nsf/44ByDocName/Green AccountingAdjustedNetSavings on January 25, 2005.

Figure 3.6
> The source Figure (S7 Fig 7.3) is based on Figure 3-9 in Intergovernmental Panel for Climate Change, 2000: *Special Report on Emissions Scenarios*, Cambridge University Press, Cambridge, U.K.

Figures 4.1 and 4.2
> The source Figures (S7 Figs 7.6a and 7.6b) are based on data downloaded from the online World Bank database and reported in World Bank, 2004: *World Development Report 2004: Making Services Work for Poor People*, World Bank, Washington D.C.

Figure 8.1
> The source Figure (C5 Box 5.2) is redrawn from Figure 7 in World Bank, 2004: *State and Trends of the Carbon Market - 2004*. World Bank, Washington D.C.

APPENDIX E
ASSESSMENT REPORT TABLES OF CONTENTS

Note that text references to CF, CWG, SWG, RWG, or SGWG refer to the entire working group report. ES refers to the Main Messages in a chapter.

Ecosystems and Human Well-being: A Framework for Assessment

CF.1	Introduction and Conceptual Framework
CF.2	Ecosystems and Their Services
CF.3	Ecosystems and Human Well-being
CF.4	Drivers of Change in Ecosystems and Their Services
CF.5	Dealing with Scale
CF.6	Concepts of Ecosystem Value and Valuation Approaches
CF.7	Analytical Approaches
CF.8	Strategic Interventions, Response Options, and Decision-making

Current State and Trends: Findings of the Condition and Trends Working Group

SDM	Summary
C.01	MA Conceptual Framework
C.02	Analytical Approaches for Assessing Ecosystem Conditions and Human Well-being
C.03	Drivers of Change (note: this is a synopsis of Scenarios Chapter 7)
C.04	Biodiversity
C.05	Ecosystem Conditions and Human Well-being
C.06	Vulnerable Peoples and Places
C.07	Fresh Water
C.08	Food
C.09	Timber, Fuel, and Fiber
C.10	New Products and Industries from Biodiversity
C.11	Biological Regulation of Ecosystem Services
C.12	Nutrient Cycling
C.13	Climate and Air Quality
C.14	Human Health: Ecosystem Regulation of Infectious Diseases
C.15	Waste Processing and Detoxification
C.16	Regulation of Natural Hazards: Floods and Fires
C.17	Cultural and Amenity Services
C.18	Marine Fisheries Systems
C.19	Coastal Systems
C.20	Inland Water Systems
C.21	Forest and Woodland Systems
C.22	Dryland Systems
C.23	Island Systems
C.24	Mountain Systems
C.25	Polar Systems
C.26	Cultivated Systems
C.27	Urban Systems
C.28	Synthesis

Scenarios: Findings of the Scenarios Working Group

SDM	Summary
S.01	MA Conceptual Framework
S.02	Global Scenarios in Historical Perspective
S.03	Ecology in Global Scenarios
S.04	State of Art in Simulating Future Changes in Ecosystem Services
S.05	Scenarios for Ecosystem Services: Rationale and Overview
S.06	Methodology for Developing the MA Scenarios
S.07	Drivers of Change in Ecosystem Condition and Services
S.08	Four Scenarios
S.09	Changes in Ecosystem Services and Their Drivers across the Scenarios
S.10	Biodiversity across Scenarios
S.11	Human Well-being across Scenarios
S.12	Interactions among Ecosystem Services
S.13	Lessons Learned for Scenario Analysis
S.14	Policy Synthesis for Key Stakeholders

Policy Responses: Findings of the Responses Working Group

SDM	Summary
R.01	MA Conceptual Framework
R.02	Typology of Responses
R.03	Assessing Responses
R.04	Recognizing Uncertainties in Evaluating Responses
R.05	Biodiversity
R.06	Food and Ecosystems
R.07	Freshwater Ecosystem Services
R.08	Wood, Fuelwood, and Non-wood Forest Products
R.09	Nutrient Management
R.10	Waste Management, Processing, and Detoxification
R.11	Flood and Storm Control
R.12	Ecosystems and Vector-borne Disease Control
R.13	Climate Change
R.14	Cultural Services
R.15	Integrated Responses
R.16	Consequences and Options for Human Health
R.17	Consequences of Responses on Human Well-being and Poverty Reduction
R.18	Choosing Responses
R.19	Implications for Achieving the Millennium Development Goals

Multiscale Assessments: Findings of the Sub-global Assessments Working Group

SDM	Summary
SG.01	MA Conceptual Framework
SG.02	Overview of the MA Sub-global Assessments
SG.03	Linking Ecosystem Services and Human Well-being
SG.04	The Multiscale Approach
SG.05	Using Multiple Knowledge Systems: Benefits and Challenges
SG.06	Assessment Process
SG.07	Drivers of Ecosystem Change
SG.08	Condition and Trends of Ecosystem Services and Biodiversity
SG.09	Responses to Ecosystem Change and their Impacts on Human Well-being
SG.10	Sub-global Scenarios
SG.11	Communities, Ecosystems, and Livelihoods
SG.12	Reflections and Lessons Learned

Millennium Ecosystem Assessment Order Form

SAVE $50 WHEN YOU PURCHASE THE FIVE-VOLUME CLOTH SET!			
Ecosystems and Human Well-being: **Five-Volume Set** *Includes Volumes 1-4 and Our Human Planet	_____ Qty. **$275.00** 1-59726-042-8	_____ Qty. **$500.00** 1-59726-041-X	

Title	Paper	Cloth	Sub-Total
Vol. 1, Current State and Trends: Findings of the Condition and Trends Working Group	_____ Qty. **$75.00** 1-55963-228-3	_____ Qty. **$145.00** 1-55963-227-5	
Vol. 2, Scenarios: Findings of the Scenarios Working Group	_____ Qty. **$65.00** 1-55963-391-3	_____ Qty. **$130.00** 1-55963-390-5	
Vol. 3, Policy Responses: Findings of the Responses Working Group	_____ Qty. **$55.00** 1-55963-270-4	_____ Qty. **$110.00** 1-55963-269-0	
Vol. 4, Multiscale Assessments: Findings of the Sub-global Assessments Working Group	_____ Qty. **$55.00** 1-55963-186-4	_____ Qty. **$110.00** 1-55963-185-6	
Our Human Planet: Summary for Decision-makers	_____ Qty. **$25.00** 1-55963-387-5	_____ Qty. **$55.00** 1-55963-386-7	
Ecosystems and Human Well-being: **Synthesis**	_____ Qty. **$15.00** 1-59726-040-1		
Ecosystems and Human Well-being: **A Framework for Assessment**	_____ Qty. **$25.00** 1-55963-403-0	_____ Qty. **$50.00** 1-55963-402-2	
		Total Book Price	
		Tax (D.C. 5.75%)	
		Shipping & Handling (U.S.: $4.50 first book; $1.00 each additional; Int'l: $5.50 first book; $1.00 each additional)	
		Total Payment	

ISLANDPRESS

Order Online: www.IslandPress.org
Call: 1-800-621-2736 · Outside of the U.S.: 773-702-7000

Name: _____

Address: _____

_____

City/State/Zip: _____

Phone: _____

E-mail: _____

□ Purchase Order No: _____

□ My check is enclosed.

Please charge my: □ Visa □ MasterCard

□ American Express □ Discover

Exp. Date: _____

Signature: _____

Mail orders to: Island Press
c/o University of Chicago Distribution Center, 11030 South Langley Ave., Chicago, IL 60628